NIST Special Publication 800-81-2

Secure Domain Name System (DNS) Deployment Guide

Ramaswamy Chandramouli
Computer Security Division
Information Technology Laboratory

Scott Rose
Advanced Network Technology Division
Information Technology Laboratory

September 2013

U.S. Department of Commerce
Penny Pritzker, Secretary

National Institute of Standards and Technology
Patrick D. Gallagher, Under Secretary of Commerce for Standards and Technology and Director

Authority

This publication has been developed by NIST to further its statutory responsibilities under the Federal Information Security Management Act (FISMA), Public Law (P.L.) 107-347. NIST is responsible for developing information security standards and guidelines, including minimum requirements for Federal information systems, but such standards and guidelines shall not apply to national security systems without the express approval of appropriate Federal officials exercising policy authority over such systems. This guideline is consistent with the requirements of the Office of Management and Budget (OMB) Circular A-130, Section 8b(3), *Securing Agency Information Systems*, as analyzed in Circular A-130, Appendix IV: *Analysis of Key Sections*. Supplemental information is provided in Circular A-130, Appendix III, *Security of Federal Automated Information Resources*.

Nothing in this publication should be taken to contradict the standards and guidelines made mandatory and binding on Federal agencies by the Secretary of Commerce under statutory authority. Nor should these guidelines be interpreted as altering or superseding the existing authorities of the Secretary of Commerce, Director of the OMB, or any other Federal official. This publication may be used by nongovernmental organizations on a voluntary basis and is not subject to copyright in the United States. Attribution would, however, be appreciated by NIST.

Certain commercial entities, equipment, or materials may be identified in this document in order to describe an experimental procedure or concept adequately. Such identification is not intended to imply recommendation or endorsement by NIST, nor is it intended to imply that the entities, materials, or equipment are necessarily the best available for the purpose.

There may be references in this publication to other publications currently under development by NIST in accordance with its assigned statutory responsibilities. The information in this publication, including concepts and methodologies, may be used by Federal agencies even before the completion of such companion publications. Thus, until each publication is completed, current requirements, guidelines, and procedures, where they exist, remain operative. For planning and transition purposes, Federal agencies may wish to closely follow the development of these new publications by NIST.

Organizations are encouraged to review all draft publications during public comment periods and provide feedback to NIST. All NIST Computer Security Division publications, other than the ones noted above, are available at http://csrc nist.gov/publications.

Reports on Computer Systems Technology

The Information Technology Laboratory (ITL) at the National Institute of Standards and Technology (NIST) promotes the U.S. economy and public welfare by providing technical leadership for the Nation's measurement and standards infrastructure. ITL develops tests, test methods, reference data, proof of concept implementations, and technical analyses to advance the development and productive use of information technology. ITL's responsibilities include the development of management, administrative, technical, and physical standards and guidelines for the cost-effective security and privacy of other than national security-related information in Federal information systems. The Special Publication 800-series reports on ITL's research, guidelines, and outreach efforts in information system security, and its collaborative activities with industry, government, and academic organizations.

Abstract

The Domain Name System (DNS) is a distributed computing system that enables access to Internet resources by user-friendly domain names rather than IP addresses, by translating domain names to IP addresses and back. The DNS infrastructure is made up of computing and communication entities called Name Servers each of which contains information about a small portion of the domain name space. The domain name data provided by DNS is intended to be available to any computer located anywhere in the Internet.This document provides deployment guidelines for securing DNS within an enterprise. Because DNS data is meant to be public, preserving the confidentiality of DNS data. The primary security goals for DNS are data integrity and source authentication, which are needed to ensure the authenticity of domain name information and maintain the integrity of domain name information in transit. This document provides extensive guidance on maintaining data integrity and performing source authentication. DNS components are often subjected to denial-of-service attacks intended to disrupt access to the resources whose domain names are handled by the attacked DNS components. This document presents guidelines for configuring DNS deployments to prevent many denial-of-service attacks that exploit vulnerabilities in various DNS components.

Keywords

Authoritative Name Server; Caching Name Server; Domain Name System (DNS); DNS Query/Response; DNS Security Extensions (DNSSEC); Resource Record (RR); Trust Anchor; Validating Resolver

Acknowledgements

The authors, Ramaswamy Chandramouli and Scott Rose of the National Institute of Standards and Technology (NIST), wish to thank their colleagues who reviewed drafts of this document. Special thanks are due for some members of Government DNSSEC working Group who provided useful feedback and pointers to some of the documents referred to in this document. We also thank Tim Grance, program manager of the Cyber and Network Security program and Doug Montgomery of the Advanced Network Technologies Division for their leadership and guidance throughout this project. Last but not the least, we are grateful to Douglas Maughan of the Department of Homeland Security for the sponsorship of this effort. The authors would also like to thank those that provided valuable feedback on the original revision of this Special Publication.

Table of Contents

List of Appendices

List of Figures

List of Tables

Executive Summary

The Internet is the world's largest computing network, with hundreds of million of users. From the perspective of a user, each node or resource on this network is identified by a unique name—the domain name—such as www.nist.gov. However, from the perspective of network equipment that routes communications across the Internet, the unique identifier for a resource is an Internet Protocol (IP) address, such as 172.30.128.27. To access Internet resources by user-friendly domain names rather than IP addresses, users need a system that translates domain names to IP addresses and back. This translation is the primary task of an engine called the Domain Name System (DNS).

The DNS infrastructure is made up of computing and communication entities that are geographically distributed throughout the world. There are more than 250 top-level domains, such as .gov and .com, and several million second-level domains, such as nist.gov and ietf.org. Accordingly, there are many name servers in the DNS infrastructure, which each contain information about a small portion of the domain name space. The DNS infrastructure functions through collaboration among the various entities involved. The domain name data provided by DNS is intended to be available to any computer located anywhere in the Internet.

This document provides deployment guidelines for securing DNS within an enterprise. Because DNS data is meant to be public, preserving the confidentiality of DNS data pertaining to publicly accessible IT resources is not a concern. The primary security goals for DNS are data integrity and source authentication, which are needed to ensure the authenticity of domain name information and maintain the integrity of domain name information in transit. This document provides extensive guidance on maintaining data integrity and performing source authentication. Availability of DNS services and data is also very important; DNS components are often subjected to denial-of-service attacks intended to disrupt access to the resources whose domain names are handled by the attacked DNS components. This document presents guidelines for configuring DNS deployments to prevent many denial-of-service attacks that exploit vulnerabilities in various DNS components.

DNS is susceptible to the same types of vulnerabilities (platform, software, and network-level) as any other distributed computing system. However, because it is an infrastructure system for the global Internet, it has the following special characteristics not found in many distributed computing systems:

- No well-defined system boundaries—participating entities are not subject to geographic or topologic confinement rules

- No need for data confidentiality—the data should be accessible to any entity regardless of the entity's location or affiliation.

Because of these characteristics, conventional network-level attacks such as masquerading and message tampering, as well as violations of the integrity of the hosted and disseminated data, have a completely different set of functional impacts, as follows:

- A masquerader that spoofs the identity of a DNS node can deny access to services for the set of Internet resources for which the node provides information (i.e., domains served by the node). This denial is not only for a limited set of clients but for the entire universe of all clients needing access to those resources

- Bogus DNS information provided by a masquerader or intruder can poison the information cache of the DNS node providing that subset of DNS information (i.e., the name server providing

Internet access service to the enterprise's users), resulting in a denial of service to the resources serviced by it

- Violation of the integrity of DNS information resident on its authoritative source or the information cache of an intermediary that has accumulated information from several historical queries may break the chained information retrieval process of DNS. This could result in either a denial of service for DNS name resolution function or misdirection of users to a harmful set of illegitimate resources.

- If the name resolution data hosted by the DNS system violates content requirements as defined in DNS standards, it could have adverse impacts such as increased workload on the DNS system, or serving obsolete data that could result in denial of service to Internet resources. In most software, program data independence (as in conventional Database Management Systems (DBMS)) provides a degree of buffer against adverse impacts due to erroneous data. In the case of DNS, the data content determines the integrity of the entire system.

Based on these functional impacts, the deployment guidelines for secure DNS presented in this document broadly consist of the following generic and DNS-specific recommendations:

- Implement appropriate system and network security controls for securing the DNS hosting environment, such as operating system and application patching, process isolation, and network fault tolerance.

- Protect DNS transactions such as update of DNS name resolution data and data replication that involve DNS nodes within an enterprise's control. The transactions should be protected using hash-based message authentication codes based on shared secrets, as outlined in Internet Engineering Task Force's (IETF) Transaction Signature (TSIG) specification.

- Protect the ubiquitous DNS query/response transaction that could involve any DNS node in the global Internet using digital signatures based on asymmetric cryptography, as outlined in IETF's Domain Name System Security Extension (DNSSEC) specification.

Enforce content control of DNS name resolution data using a set of integrity constraints that are able to provide the right balance between performance and integrity of the DNS system. This guide contains recommendations for securing a DNS name server. Part of those recommendations is deployment of the DNS Security Extensions (DNSSEC) for zone information. The basic steps to accomplish that part of security are below:

- Install a DNSSEC capable name server implementation (See Section 7.2.1)

- Check zone file(s) for any possible integrity errors (See Section 10)

- Generate asymmetric key pair for each zone and include them in the zone file (See Section 9.2 and 9.3)

- Sign the zone (See Section 9.6)

- Load the signed zone onto the server

- Configure name server to turn on DNSSEC processing (See Section 9.1)

- (OPTIONAL) send copy of public key to parent for secure delegation

Note that this guide focuses on authoritative name servers for the most part, but the basic steps of DNSSEC deployment for caching name servers are below:

- Install a DNSSEC capable resolver implementation (See Section 7.2.1)

- Obtain one or more trust anchors for zones administrator wants validated (See Section 9.7)

- Configure resolver to turn on DNSSEC processing (See Section 9.1)

The rest of the guide covers recommendations for secure configuration and operations of name servers.

Changes in this Document

The following changes appear in this revision over the initial release of NIST Special Publication 800-81:

- Updated recommendations for cryptographic parameters based on previous NIST Special Publication 800-57 [800-57P1].

- Included discussion of NSEC3 Resource Records in DNSSEC

- Discussion of DNSSEC in split view deployments

- Minor fixes of examples and text

- Inclusion of examples based on NSD as well as BIND software packages

1. Introduction

1.1 Authority

The National Institute of Standards and Technology (NIST) developed this document in furtherance of its statutory responsibilities under the Federal Information Security Management Act (FISMA) of 2002, Public Law 107-347.

NIST is responsible for developing standards and guidelines, including minimum requirements, for providing adequate information security for all agency operations and assets; but such standards and guidelines shall not apply to national security systems. This guideline is consistent with the requirements of the Office of Management and Budget (OMB) Circular A-130, Section 8b(3), "Securing Agency Information Systems," as analyzed in A-130, Appendix IV: Analysis of Key Sections. Supplemental information is provided in A-130, Appendix III.

This guideline has been prepared for use by Federal agencies. It may be used by nongovernmental organizations on a voluntary basis and is not subject to copyright, though attribution is desired.

Nothing in this document should be taken to contradict standards and guidelines made mandatory and binding on Federal agencies by the Secretary of Commerce under statutory authority, nor should these guidelines be interpreted as altering or superseding the existing authorities of the Secretary of Commerce, Director of the OMB, or any other Federal official.

1.2 Purpose and Scope

This publication seeks to assist organizations in understanding the secure deployment of Domain Name System (DNS) services in an enterprise. It provides practical, real-world guidance on securing each facet of DNS within an organization based on an analysis of the operating environment and associated threats.

Currently, the DNS is not the target of most attacks, but as hosts become more security aware, and applications begin to rely on the DNS infrastructure for network operations, the DNS infrastructure will become a more tempting target. The ultimate goal for DNSSEC is full deployment across the entire domain tree on the infrastructure side, and implementation in applications that can demand the services provided by DNSSEC. At present there are no operational nodes in the DNS domain tree that provides DNSSEC capabilities. Hence the first step towards fully deployment is to provide DNSSEC capability for domain subtrees that have high security needs. Once DNSSEC capabilities become widely available in the infrastructure, application developers will be able to develop DNSSEC-aware applications and thus use DNSSEC as a means for network security.

In this guide, DNSSEC deployment is targeting the DNS infrastructure, not individual hosts. However, as the infrastructure becomes more secure, DNSSEC will naturally push down to the individual clients that make DNS queries. DNSSEC was designed with backward compatibility in mind, so that current network applications can gain some security from DNSSEC, provided that servers upstream are using DNSSEC, but in the future, it is hoped all systems (DNS name servers and clients) will be able to perform at least some of the operations detailed in the DNSSEC specifications and this document.

1.3 Audience

This document has been created for the administrators of DNS deployments, as well as computer security staff and system administrators who are responsible for performing duties related to DNS.

1.4 Document Structure

The remainder of this document is organized into the following ten major sections:

- Section 2 provides an introduction to DNS and DNS infrastructures. It also discusses the security objectives of DNS and provides an overview of the aspects of DNS addressed by this document.

- Section 3 introduces some basic DNS components, such as the zone file that holds DNS data, and the name servers and resolvers that provide DNS services.

- Section 4 defines the different types of DNS transactions.

- Section 5 discusses the threats, security objectives, and protection approaches involving the DNS hosting environment. Section 6 provides the same types of information for DNS transactions.

- Sections 7 and 8 present guidelines for securing the DNS hosting environment and DNS transactions (except DNS query/response), respectively.

- Section 9 provides recommendations for securing DNS query/response Transaction.

- Section 10 focuses on guidance for minimizing information exposure through DNS.

- Section 11 presents guidelines for DNS security administration operations.

The document also contains appendices with supporting material. Appendix A presents definitions of important DNS security terminology. Appendix B gives information for vendor specific information. Appendix C contains an acronym list, and Appendix D contains a bibliography.

2.   Securing Domain Name System

This document provides deployment guidelines for securing the Domain Name System (DNS) in an organization. The deployment guidelines follow from an analysis of security objectives and consequent protection approaches for all DNS components. The rationale for security objectives and mechanics for development of deployment guidelines are given below:

- Security objectives for each DNS component are developed on the basis of analysis of operating environment and associated threats.

- Secure deployment guidelines for each DNS component are provided through a combination of configuration options and checklists that are based on policies or best practices.

The primary security specifications (with associated mechanisms) for which this document provides deployment guidelines are as follows:

- Internet Engineering Task Force (IETF) Domain Name System Security Extensions (DNSSEC) specifications, covered by Request for Comments (RFC) 4033, 4034, 4035, and 3833 [RFC4033], [RFC4034], [RFC4035], [RFC3833].

- IETF Transaction Signature (TSIG) specifications, covered by RFCs 2845 and 3007 [RFC2845], [RFC3007].

2.1   What is the Domain Name System (DNS)?

The Internet is the world's largest computing network, with more than 580 million users. From the perspective of a user, each node or resource on this network is identified by a unique name: the domain name. Some examples of Internet resources are:

- Web servers—for accessing Web sites

- Mail servers—for delivering e-mail messages

- Application servers—for accessing software systems and databases remotely.

From the perspective of network equipment (e.g., routers) that routes communication packets across the Internet, however, the unique resource identifier is the Internet Protocol (IPv4 or IPv6) address, represented as a series of four numbers separated by dots (e.g., 123.67.43.254). To access Internet resources by user-friendly domain names rather than these IP addresses, users need a system that translates these domain names to IP addresses and back. This translation is the primary task of an engine called the *Domain Name System* (DNS).

Users access an Internet resource (e.g., a Web server) through the corresponding client or user program (e.g., a Web browser) by typing the domain name. To contact the Web server and retrieve the appropriate Web page, the browser needs the corresponding IP address. It calls DNS to provide this information. This function of mapping domain names to IP addresses is called *name resolution*. The protocol that DNS uses to perform the name resolution function is called the DNS protocol.

The DNS function described above includes the following building blocks. First, DNS should have a data repository to store the domain names and their associated IP addresses. Because the number of domain names is large, scalability and performance considerations dictate that it should be distributed. The domain names may even need to be replicated to provide fault tolerance. Second, there should be software that manages this repository and provides the name resolution function. These two functions (managing the domain names repository and providing name resolution service) are provided by the primary DNS component, the *name server*. There are many categories of name servers, distinguished by type of data served and functions performed. To access the services provided by a DNS name server on behalf of user programs, there is another component of DNS called the *resolver*. There are two primary categories of resolvers (caching/recursive/resolving name server and stub resolver),[1] distinguished by functionality. The communication protocol; the various DNS components; the policies governing the configuration of these components; and procedures for creation, storage, and usage of domain names constitute the DNS infrastructure.

2.2   DNS Infrastructure

The DNS infrastructure is made up of computing and communication entities that are geographically distributed throughout the world. To understand this DNS infrastructure, it is necessary to examine first the structure behind the organization of domain names. The domain name space (the universe of all domain names) is organized in the form of a hierarchy. The topmost level in the hierarchy is the *root domain*, which is represented as a dot ("."). The next level in the hierarchy is called the *top-level domain* (TLD). There is only one root domain, but there are many TLDs. Each TLD is called a child domain of the root domain. In this context, the root domain is the parent domain because it is one level above a TLD. Each TLD, in turn, can have many child domains. The children of TLDs are called *second-level* or *enterprise-level domains*.

In a domain name representation, the symbol for the root domain usually is omitted. For example, consider the domain name marketing.example.com. The rightmost label in this domain name ("com.") is a TLD. The next label to the left ("example") is the second-level or enterprise-level domain. The leftmost label ("marketing") is the third-level domain. It also is possible to have a fourth-level domain, fifth-level domain, and so forth. Because each of the labels in marketing.example.com is called a domain (TLD, second-level domain, third-level domain, etc.), the concatenation of all these labels from the current level to the TLD is a *fully qualified domain name* (FQDN). In this document, however, the FQDN is referred to as simply a domain name, and the level name is used to identify individual labels.

There is only one root domain. There are several hundreds (possibly soon to be thousands) of TLD's, categorized into the following three types:

- **Country-code TLDs (ccTLDs)**—domains associated with countries and territories. There are more than 240 ccTLDs. Examples include .uk, .in, and .jp.

- **Sponsored generic TLDs (gTLDs)**—specialized domains with a sponsor representing a community of interest. These TLDs include .edu, .gov, .int, .mil, .aero, .coop, and .museum.

- **Unsponsored generic TLDs (gTLDs)**—domains without a sponsoring organization. The list of unsponsored gTLDs includes .com, .net, .org, .biz, .info, .name, and .pro.

There are several million enterprise-level (second-level or lower) domains. In fact, as of December 2008 there were more than 77 million registered domain names in the .com gTLD alone. A partial DNS name space hierarchy is shown in Figure 2-1.

---

[1]   The caching/recursive/resolving name server is included under the category of resolver since it plays the dual role of being a resolver and a name server.

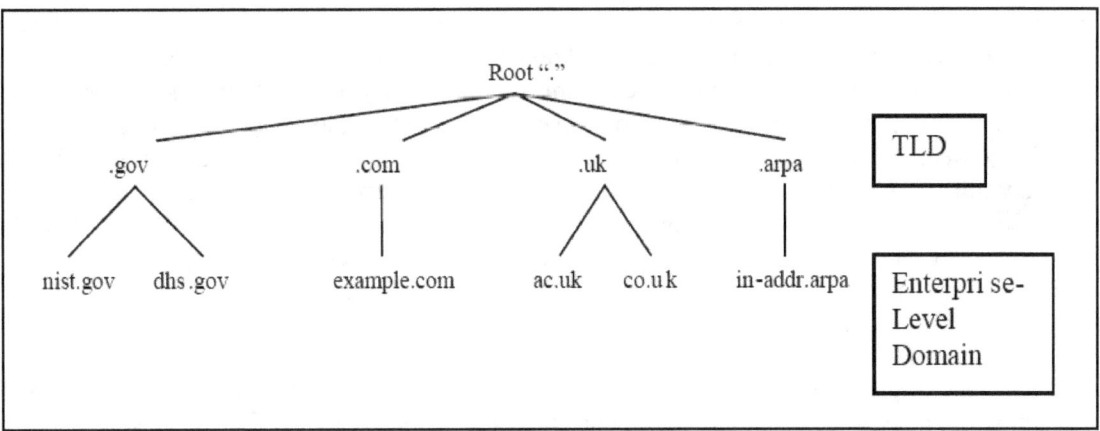

Figure 2-1. Partial DNS Name Space Hierarchy

There are many name servers in the DNS infrastructure. Each name server contains information about a portion of the domain name space. Name servers are associated with levels as far as the first three levels of domain name space are concerned. There are 13 name servers associated with the root level; they are called *root servers*. Two of the root servers are currently run by the U.S private-sector corporation VeriSign; the rest are operated by other organizations around the world as a service to the Internet community. The organizations that run name servers associated with a TLD are called *registries*. Generally, ccTLDs are run by designated registries in the respective countries, and global registries run gTLDs. For example, VeriSign currently manages the name servers for the .com and .net TLDs, a nonprofit entity called Public Internet Registry (PIR) manages the name servers for the .org TLD, and another nonprofit organization called EDUCAUSE manages the name servers for the .edu TLD. All of these registry organizations are subject to change, however. Domain registrants should be familiar with the points of contact (called registrars) for their particular registry. The name servers associated with enterprise-level domains and below are either run directly by the organizations that own those domains or outsourced to Internet service providers (ISP) or other service providers.

The DNS infrastructure functions through collaboration among the various entities involved (organizations that manage root servers, registries that run TLDs, etc.) A nonprofit, private-sector corporation, the Internet Corporation for Assigned Names and Numbers (ICANN), acts as the technical coordination body for aspects of DNS. For example, ICANN formulates policies for management of root servers. ICANN also is the authority for creation of new TLDs. ICANN was established in 1998 by the U.S. Department of Commerce.

A user (i.e., an individual or corporation) wishing to register a domain name (which can only be an enterprise-level domain under a TLD) must contact an authorized entity called a registrar (which may charge a fee, depending on the TLD in question). *Registrars* are companies that are authorized to register domain names in a particular TLD (or even in a sub-domain of a TLD e.g. co.uk.) to end-users. There are registrars all over the world. When the registrar receives the user's registration request, the registrar verifies that the name is available by checking with the registry that manages the corresponding TLD (or sub-domain under the TLD). If the name is available, the registrar registers the name with the appropriate registry. The registry then adds the new name to its registry database and publishes the new name in DNS. Note that in some domains (some country codes and gTLDs for example), the same organization acts as the registry and registrar for names in the domain. There is no middle organization that registers domain names on behalf of the domain holder.

Organizations that register and obtain an enterprise-level domain name often have to create child domains to properly identify Internet resources associated with various functional units. For example, the owner of the domain name example.com might create the subdomain shipping to create and identify resources associated with the shipping department of the organization. Similarly, many other subdomains (in this context, third-level domains) may be created to properly identify all of the Internet resources of the organization. Often, however, in any one organization (that is, the owner of a second-level domain) there will be many third-level domains but few Internet resources (Web servers, application servers, etc.) in each of these domains. Hence, it is not economical to assign a unique name server for each of these third-level and lower-level domains. Furthermore, it is administratively convenient to group all information pertaining to an organization's primary domain (i.e., a second-level or enterprise-level domain) and all its subdomains into a single resource and manage it as a unit.

To facilitate this grouping, the DNS defines the concept of a zone. A zone may be either an entire domain or a domain with one or more subdomains. A zone is a configurable entity within a name server under which information on all Internet resources pertaining to a domain and a selected set of subdomains is described. Thus, zones are administrative building blocks of the DNS name space just as domains are the structural building blocks. As a result, the term *zone* commonly is used even to refer to a domain that is managed as a standalone administrative entity (e.g., the root zone, the .com zone). This document uses the term *zone* to refer to a resource entity that contains domain name information about one or more domains and is managed as a single administrative entity. In other words, the zone is the configurable resource inside a deployed name server installation that contains the domain name information.

With this overall knowledge of DNS infrastructure (name servers and resolvers), domain names, zones, name servers of various levels (i.e., root servers and TLD servers) and resolvers, the name resolution function can now be defined in more detail. When a user types the URL www.example.com into a Web browser, the browser program contacts a type of resolver called a *stub resolver* that then contacts a local name server (called a recursive name server or resolving name server). The resolving name server will check its cache to determine whether it has valid information (the information is determined to be valid on the basis of criteria described later in this document) to provide IP address for the accessed Internet resource (i.e., www.marketing.example.com). If not, the resolving name server checks the cache to determine whether it has the information regarding the name server for the zone marketing.example.com (since this is the zone that is expected to contain the resource www.marketing.example.com). If the name server's IP address is in the cache, the resolver's next query will be directed against that name server. If the IP address of the name server of marketing.example.com is not available in the cache, the resolver determines whether it has the name server information for a zone that is one level higher than marketing.example.com (i.e., example.com). If the name server information for example.com is not available, the next search will be for the name server of the .com zone in the cache.

If the complete search of the cache (as described above) does not yield the required information, the resolving name server has no alternative but to start its search by querying the name server in the topmost zone in the DNS name space hierarchy (i.e., the root server). If the cache search is successful, the resolving name server has to query the name servers in one of the levels below the root zone (in this context, either marketing.example.com, example.com, or .com). Because the set of iterative queries starting from the root server subsumes the set starting from any of the lower-level servers, this section describes the name resolution process starting from the query sent to the root server by the resolving name server at the enterprise-level.

Contact with the root servers is enabled by a file called the "root hints" file that is usually present in every name server in DNS. The root hints file contains the IP address of one or more of the 13 root servers. The root server will contain information about the name servers for its child zones (i.e., TLDs). A TLD (e.g., .com) will contain name server information about its child zones (e.g., example.com). The name server information about its child zones that is carried in a zone is called delegation information. The delegation information is the one that is used by a zone to refer name resolution requests for a resource lower than it in the domain name hierarchy. Since the name resolution request in this example pertains to a resource in the third-level domain, the root server must refer the request to a lower-level name server. The response to the resolving name server that involves sending this delegation information is called the referral. The *referral* provides the name and IP address for the name servers for the TLD zone that is relevant to the request (i.e., the .com zone) (since the query is for a resource in marketing.example.com). Using this referral, the resolving name server then formulates and sends a query to the .com zone name server. This server will provide the referral for example.com's name server. If the marketing.example.com domain is included in example.com's zone, querying the name server for example.com will provide the IP address for the resource www.marketing.example.com. A diagram of the name resolution process (without cache search operations) is given in Figure 2-2.

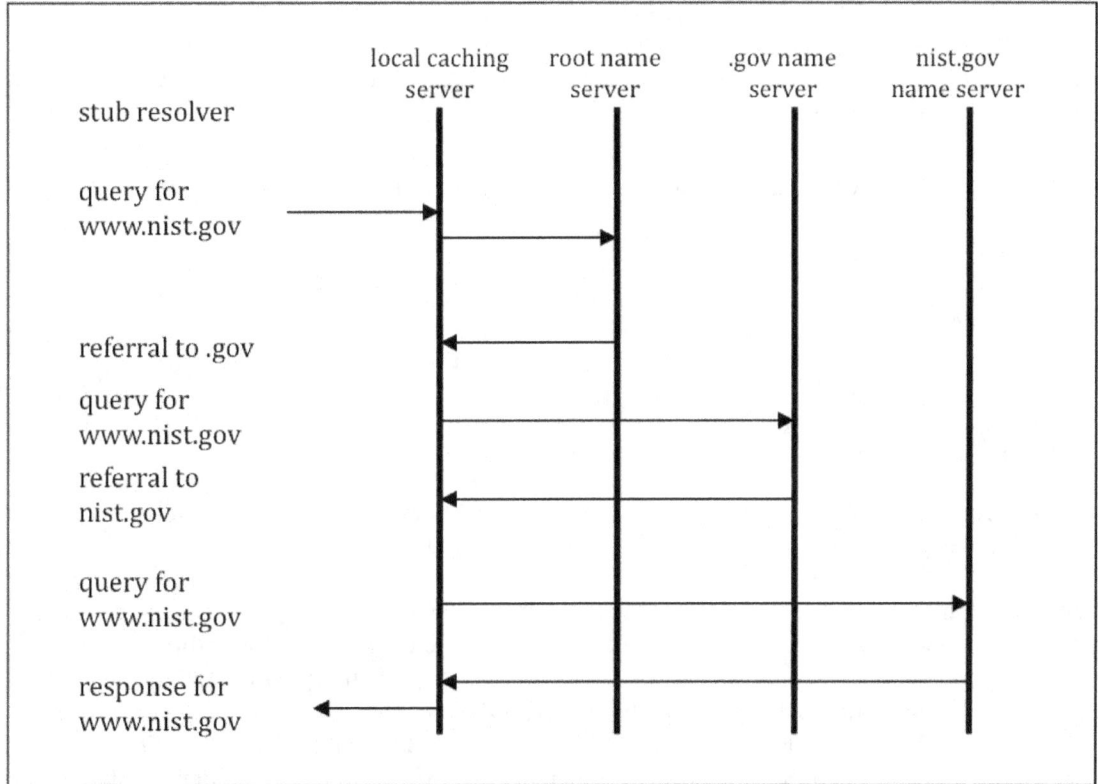

Figure 2-2. Name Resolution Process (without cache search)

As the description of the name resolution process makes clear, a name server performs the following functions:

- It provides a referral to a child zone.

- It provides a mapping from domain name to IP address, called *domain name resolution*, or IP address to domain name, sometimes called an *inverse resolution*, but is actually a standard query for a different type of data.

- It provides an error message if the query is for a DNS entry that does not exist.

The name server performs these three functions with the same DNS database, which is called a *zone file*. A class of information called *delegation information* contains the name server information for child zones in a parent zone and is used in performing the referral function. The mapping function is performed by a class of information in a zone file called *authoritative information* and is provided by a set of records that list the resources in that zone, along with its domain name and its corresponding IP address. Because the resources belong to that zone, the information provided is deemed authoritative. Thus, a zone file contains two categories of information: authoritative information (information about all resources for all domains in the zone) and delegation information (information about name servers for child zones). The locations in the zone file where delegation information appears are called *delegation points*. The level of a zone file is the level of the topmost domain for which it contains authoritative information. In the previous example, the zone file in the name server of example.com is the enterprise-level zone file, and the corresponding name server is called the enterprise-level name server.

2.3   DNS Components and Security Objectives

Before security objectives can be determined, the building blocks of the DNS need to be specified. DNS includes the following entities:

- The platform (hardware and operating system) on which the name server and resolvers reside

- The name server and resolver software

- DNS transactions

- DNS database (zone files)

- Configuration files in the name server and resolver.

The media access-level network technology (i.e., Ethernet network connecting a stub resolver and the local resolving name server) is not included in the definition of DNS.

Confidentiality, integrity, availability, and source authentication are security goals that are common to any electronic system. However, the DNS is expected to provide name resolution information for any publicly available Internet resource. Hence except for DNS data pertaining to internal resources (e.g., servers inside a firewall etc), that is provided by internal DNS name servers through secure channels, the DNS data provided by public DNS name servers is not deemed confidential. Hence confidentiality is not one of the security goals of DNS. Ensuring the authenticity of information and maintaining the integrity of information in transit is critical for efficient functioning of the Internet, for which the DNS provides the name resolution service. Hence, integrity and source authentication are the primary DNS security goals.

Because of the sheer diversity of name server platforms and the underlying networks in which DNS components (such as name server software and resolver software) reside, preventing all types of denial-of-service attacks is not feasible. Some baseline guidelines must be observed, however, to prevent denial-of-service attacks that exploit vulnerabilities in the name server platform, zone file data content, and access control configuration for certain DNS transactions.

2.4   Focus of the Document

There are three main levels of name servers: root name servers, TLD name servers, and enterprise-level name servers. This document provides deployment guidelines for securing name servers for the .gov domain (zone) at the TLD level and for all enterprise-level name servers below the .gov TLD. Thus, the guidelines cover secure configuration and operation of all name servers in all civilian agencies of the U.S. Federal Government. The target audience consists of zone (and to some extent, system) administrators who are responsible for the configuration and operation of these name servers. The guidelines are equally applicable, however, to any enterprise-level zone (e.g., mit.edu).

The security mechanisms to which the guidelines in this document apply conform to IETF's DNSSEC and TSIG specifications. The guidelines in this document cover policies, configuration options, and checklists for the following DNS components/associated operations):

- DNS hosting environment

  - Host platform (O/S, file system, communication stack)

  - DNS software (name server, resolver)

  - DNS data (zone file, configuration file)

- DNS transactions

  - DNS query/response

  - Zone transfers

  - Dynamic updates

  - DNS NOTIFY

- Security administration

  - Choice of algorithms and key sizes (TSIG and DNSSEC)

  - Key management (generation, storage, and usage)

  - Public key publishing and setting up trust anchors

  - Key rollovers (scheduled and emergency).

3. DNS Data and DNS Software

The two primary software components of DNS are the name server and the resolver. The primary functions of the name server are to host the database (called the zone file) containing domain information and to provide responses to name resolution queries through authoritative responses or referrals. The primary function of the resolver software is to formulate a name resolution query or series of queries. The primary DNS data is the zone file (the configuration file is another type of DNS data). Section 3.1 discusses the composition of the zone file, and Sections 3.2 and 3.3 address the functions of various types of name servers and resolvers, respectively. The discussion of name servers and resolvers is in the enterprise-level context and may not apply to corresponding entities in the root and TLD levels.

3.1   DNS Zone Data

The zone file contains information about various resources in that zone. The information about each resource is represented in a record called a Resource Record (RR). Because a zone may contain several domains and several types of resources within each domain, the format of each RR contains fields for making this identification. Specifically, the RR consists of the following major fields:

- **Owner name:** the domain name or a resource name

- **TTL:** time to live in seconds

- **Class:** at present only one class, IN (denoting Internet), is used

- **RRType:** type of resource

- **RData:** information about the resource (depends upon RRType)

Some of the common RRTypes are:

- **A:** Address RRType. An RR of this type provides the IP address for a host name (identified using a FQDN).

- **MX:** Mail Exchanger RRType. An RR of this type provides the mail server host name for a domain.

- **NS:** Name Server RRType. An RR of this type provides a name server host name for a domain.

IETF RFC 1035 provides the complete format of valid RRTypes in DNS [RFC1035]. Because there could be multiple resources of a given RRType (e.g., several hosts acting as name servers) under the same owner name and since there is only one class (CLASS) (i.e., IN), the information of interest (e.g., all mail servers (RRType = MX) in a domain) in a zone file is on a combination of RRs that contain the same owner name, TTL, class, and RRType. The set of RRs with the same owner name, class, and RRType is called an RRSet. Hence, logically a zone file is made up of several RRSets.

3.2   Name Servers (DNS Software)

There are two main types of name servers: authoritative name servers and caching name servers. The term *authoritative* is with respect to a zone. If a name server is an authoritative source for RRs for a particular zone (or zones), it is called an *authoritative name server* for that zone (or zones). An authoritative name server for a zone provides responses to name resolution queries for resources for that zone, using the RRs in its own zone file. A *caching name server* (also called a resolving/recursive name server), by contrast, provides responses either through a series of queries to authoritative name servers in the hierarchy of domains found in the name resolution query or from a cache of responses built by using previous queries.

3.2.1 Authoritative Name Servers

There are two types of authoritative name servers: master (or primary) name server and slave (or secondary) name server. To improve fault tolerance, there could be several slave name servers in an enterprise. A *master name server* contains zone files that are created and edited manually by the zone administrator. Sometimes a master name server allows the zone file to be dynamically updated by authorized DNS clients. A master name server that is configured with this feature usually is called the *primary master name server*. A *slave (secondary) name server* also contains authoritative information for a zone, but its zone file is a replication of the one in the associated master name server. The replication is enabled through a transaction called *zone transfer* that transfers all RRs from the zone file of a master name server to the slave name server. Because a name resolution query is for a specific RR, a zone transfer actually is treated as a category of name resolution query with type code AXFR, which means "all RRs for the zone that is being queried." Whenever the contents of a zone file are changed in the master name servers, the slave name servers are notified of the change through a transaction called *DNS NOTIFY*. When a slave name server receives this request, it initiates a zone transfer request to the master name server.

3.2.2 Caching Name Servers

A caching name server generally is the local name server in the enterprise that performs the name resolution function on behalf of the various enterprise clients. A caching name server also is called a resolving/recursive name server. The name resolution function is performed by a caching name server in response to queries from a stub resolver. The search process involved in name resolution may involve searching its own cache, recursively querying various authoritative name servers through a set of iterative queries, or a combination of these methods.

A specific name server can be configured to be both an authoritative and a recursive name server. In this configuration, the same name server provides authoritative information for queries pertaining to authoritative zones while it performs the resolving functions for queries pertaining to other zones. To perform the resolving function, it has to support recursive queries. Any server that supports recursive queries is more vulnerable to attack than a server that does not support such queries (see following sections). As a result, authoritative information might be compromised. Hence, it is not a good security practice to configure a single name server to perform both authoritative and recursive functions.

3.3 Resolvers (DNS Software)

Software such as Web browsers and e-mail clients that require access to Internet resources make use of the DNS client, called the *client resolver* or *stub resolver*. The stub resolver formulates a name resolution query for the resource sought by the Internet-accessing software and sends it to a caching (resolving) name server in the enterprise. Stub resolvers generally are configured to send queries to two or more resolving name servers to provide some fault tolerance for their operation. A stub resolver often is referred to as simply a *resolver*. A caching (resolving) name server that receives a query from a stub resolver also formulates queries for sending them to authoritative name servers (if it is not able to respond to the query from its cache) and hence also sometimes is referred to as a *resolver* because it has a resolving component and a caching (name server) component.

A resolver (either stub or part of a caching name server) that performs DNSSEC validation is called a validating resolver (sometimes validator for short). A validating resolver performs all the same DNS operations (i.e., formulating name resolution queries) as a stub/caching name server but also attempts to validate the responses in the form of DNSSEC signed data when available. In order for a validator to perform its validation, it must first be configured with one or more public keys as trust anchors. See Section 13 for more information about DNSSEC validation.

4. DNS Transactions

The most common types of DNS transactions are the following:

- DNS query/response

- Zone transfer

- Dynamic updates

- DNS NOTIFY.

This section describes each of these transaction types.

4.1 DNS Query/Response

This is the most common transaction in DNS. A DNS query originates from a resolver; the destination is an authoritative or caching name server. The most common query is a lookup for an RR, based on its owner name or RRType. The response may consist of a single RR, an RRSet, or an appropriate error message.

As discussed previously, there are two types of queries: iterative and recursive. DNS resolvers that send iterative queries tend to be more robust with regard to the types of responses they provide because they may have to follow multiple referrals to obtain the final answer to a given query. If they also have a DNS cache, they can build up a global view of name servers in the DNS and the responses from previous queries. This can be used to shorten the turnaround time for future queries. Recursive queries usually are sent from stub resolvers that do not have the capability to handle complex DNS operations. Instead, they rely on an upstream DNS entity (usually a name server with a cache that sends iterative queries on behalf of a collection of stub resolvers) to perform the query and return the ultimate answer.

DNS queries are sent in a single UDP packet. The response usually is a single UDP packet as well, but data size may result in truncation—in which case the normal procedure is to reissue the query using TCP. UDP is preferred because of its lower overhead in consuming resources, and DNS administrators should make sure the zone data in responses do not result in a large percentage of truncated DNS responses.

DNS queries are sent in the clear, and it is assumed that the response received is correct and from an authentic source. As a result, it is possible for an active attacker to intercept (and alter) or forge responses back to a querying client. Section 6.1 provides a more detailed examination of threats to the DNS query/response transaction. Sections 8.1.1 and 9 provide security solutions proposed to address these threats.

4.2 Zone Transfer

A *zone transfer* refers to the way a (secondary) slave server refreshes the entire contents of its zone file from the (primary) master servers. This process enables a secondary name server to keep its zone file in synch with its primary name server. A zone transfer transaction starts as a query from a secondary name server to a primary name server. A zone transfer query—in contrast to a DNS Query—requests all RRs from a zone instead of requesting RRs of a given owner name or RRType. A zone transfer query originates from a secondary name server either in response to a DNS NOTIFY message (see Section 4.4) or on the basis of the value of the Refresh data item in the RData field of the zone's Start of Authority (SOA) RR.

A zone transfer process has different security implications because it reveals a lot more information than a normal DNS Query and because of the increased resource (bandwidth and response time) usage of the message. The threats to a zone transfer transaction are discussed in Section 6.2. The protection mechanisms are discussed in sections 8.1.2 and 8.2.5.

4.3 Dynamic Updates

As enterprises add, delete, and move around IP-network-based resources (i.e., database servers, Web servers, mail servers, and even name servers), corresponding changes may need to be made to the zone file that carries information about the domains where the resources are located. In the early days of the DNS, DNS zone administrators made such changes manually. When such changes became larger in volume and more frequent, however, the manual update process was found to be inadequate, and the concept of dynamic updates was introduced. Apart from volume and frequency, there are some applications that require instant automatic updates to the DNS zone file through application programs. Examples of these applications include Certificate Authority (CA) servers, Dynamic Host Configuration Protocol (DHCP) servers, and Internet multicast address servers. Dynamic Updates are always made to the zone file of the primary (master) name server.

In a few instances, the DNS server is used as a repository for public key certificates. In these instances, a CA server receives a public key from a user, signs it with its private key to generate a certificate, and stores this certificate as a CERT or TLSA [DANE] RR in a DNS server. A DHCP server dynamically assigns an IP address to a requesting host, and then adds this information (host and newly assigned IP address) as an A-type RR in the DNS authoritative name server. The DHCP server also deletes this A-type RR after the IP address is returned by the host. An Internet multicast server selects a new multicast IP address from the Internet multicast IP address space and assigns it to the newly generated multicast group. Then the server adds this information (domain name of multicast group, newly selected multicast address) as an A-type RR in the DNS database so that users can join the multicast group using a user friendly domain name rather than an IP address.

RFC 2136 [RFC2136] outlines the roadmap for the dynamic update mechanism, which subsequently was implemented in BIND 8 version and has continued in all versions since.

The dynamic update facility provides operations for addition and deletion of RRs in the zone file. Because updates to the contents of an existing RR can be accomplished through deletion of the old record and addition of the new record with the changed contents, no separate update operation is provided.

The suite of operations included in the dynamic update facility consists of the following:

- Add or delete individual RRs for an existing domain

- Delete specific RRsets (a set of resource records with the same owner name, class, and RRType [e.g., a set of RRs with RRType NS for the domain/owner name example.com with the common class IN of course]) for an existing domain

- Delete an existing domain (all resource records for a given domain name [e.g., all RRs for the domain example.com])

- Add a new domain (one or more RRs for a new domain [e.g., adding an A-type RR for a new domain NYBranch.example.com]).

DNS zones that allow dynamic update open themselves to a host of malicious attacks. The full list of potential attacks is detailed in Section 6.3. The proposed solution to limit or prevent these attacks is given in sections 8.1.4, 8.2.7, 8.2.8 and 8.2.9.

4.4   DNS NOTIFY

Whenever changes occur in the zone file of the primary (master) DNS server, the secondary (slave) DNS servers that are supposed to carry identical data as the primary DNS server must be notified of the changes. This notification is accomplished through the *DNS NOTIFY message*, which signals a secondary DNS server to initiate a zone transfer (see Section 4.2). The DNS NOTIFY message is a much more efficient and faster way of keeping secondary servers in sync with the primary server than the alternative approach of secondary servers polling the primary server for changes via the SOA Refresh value timeout.

Sending the DNS NOTIFY message, called a *notify operation*, is a default operation in BIND 9.x. The next question that arises is: How does the DNS name server software know to which servers the DNS NOTIFY message must be sent? The default in BIND 9.x is to notify servers that are defined in the NS RRs for the zone. If there are any additional servers to which the zone administrator wants the DNS NOTIFY message to be sent (e.g., stealth slave server), the DNS administrator can add the IP addresses of the other entities in the BIND configuration file. There are configuration options to stop the server from sending DNS NOTIFY messages to a particular zone or to all zones served by this name server.

Once a secondary server receives a DNS NOTIFY message, it resets the relevant zone's refresh value to zero, prompting a zone transfer attempt. As in any zone refresh, if the zone serial number in the SOA RR has not increased (see Section 10.1), the zone transfer does not take place. This procedure allows changes to the zone to propagate to all name servers more quickly.

Since a DNS NOTIFY message triggers zone transfer, spurious DNS NOTIFY messages could result in unnecessary zone transfers and hence potential denial of service. The proposed solution for minimizing these spurious notifications is given in Section 8.1.5 and 8.1.6.

5. DNS Hosting Environment—Threats, Security Objectives, and Protection Approaches

The DNS hosting environment consists of the following elements:

- Host platform (operating system [OS], file system, communication stack)

- DNS software (name server, resolver)

- DNS data (zone file, configuration file)

This section describes threats and recommended protection approaches for these portions of the hosting environment.

5.1 Host Platform Threats

Threats to the platform that hosts DNS software are no different from threats that any host in the Internet faces. These generic threats and their impact—viewed specifically from the point of view of DNS hosts—are as follows:

- **Threat T1:** The OS, any system software, or any other application software on the DNS host could be vulnerable to attacks such as buffer overflows, resulting in denial of name resolution service.

- **Threat T2:** The TCP/IP stack in DNS hosts (stub resolver, caching/resolving/recursive name server, authoritative name server, etc.) could be subjected to packet flooding attacks (such as SYNC and smurf), resulting in disruption of communication. An application layer counterpart of this attack is to send a large number of forged DNS queries to overwhelm an authoritative or resolving name server.

- **Threat T3:** A malicious insider who has access to local area network (LAN) segments where DNS hosts reside could launch an Address Resolution Protocol (ARP) spoofing attack that disrupts DNS message flows.[2]

- **Threat T4:** The platform-level configuration file that enables communication (e.g., resolv.conf and host.conf in Unix platforms) can be corrupted by viruses and worms or subject to unauthorized modifications due to inadequate file-level protections, resulting in breakdown of communication among DNS hosts (e.g., between a stub resolver and a resolving name server, between a resolving name server and an authoritative name server).

- **Threat T5:** The DNS-specific configuration files (named.conf, root.hints, etc.), data files (zone file), and files containing cryptographic keys could be corrupted by viruses and worms or subjected to unauthorized modifications due to inadequate file-level protections, resulting in improper functioning of name resolution service.

---

[2] This is not strictly a host threat, but rather a network threat, which is mitigated by placing DNS servers within their own restricted LAN segments (e.g., via VLANs). Since generic network level threats are outside the scope of this document, this threat has been included since it involves a DNS parameter (i.e., IP address).

- **Threat T6:** A malicious host on the same LAN as a DNS client may be able to intercept and/or alter DNS responses. This would allow an attacker to redirect a client to a different site. This could be the first action in an attack on a client host.

5.2   DNS Software Threats

Threats to the DNS software itself can have serious security impacts. The most common software problems and the impact of threats against them are as follows:

- **Threat T7:** DNS software (name server or resolver) could have vulnerabilities such as buffer overflows that result in denial of service.

- **Threat T8:** DNS software does not provide adequate access control capabilities for its configuration files (e.g., named.conf), its data files (e.g., zone file) and files containing signing keys (e.g., TSIG, DNSKEY) to prevent unauthorized read/update of these files. These capabilities are provided on top of O/S-file level protection referred to in threats T4 and T5 and may depend upon the latter.

5.3   Threats Due to DNS Data Contents

DNS data is made up of two types: zone files and configuration files. The content of both these types of DNS data has security ramifications. All the security deployment options discussed in this document relate to configuration file contents. Security implications due to zone file content are discussed in the section titled "Guidelines for Minimizing Information Exposure through DNS Content Control" (Section 10) and are mostly due to the following aspects of zone data:

- Parameter values for certain key fields in RRs of various RRTypes

- Presence of certain RRs in the zone file.

The various types of undesirable contents in the zone file results in different security exposures and consequent potential threats as follows:

- **Threat T9—Lame Delegation:** This error occurs when FQDN and/or IP addresses of name servers have been changed in the child zone but the parent zone has not updated the delegation information (NS RRs and glue records). In this situation, the child zone becomes unreachable (denial of service).

- **Threat T10—Zone Drift and Zone Thrash:** If the Refresh, and Retry, fields in the SOA RR of the primary name server are set too high and the zone file is changed frequently, there may be a mismatch of data between the primary and secondary name servers. This error is called *zone drift*; it results in incorrect zone data at the secondary name servers. If the Refresh and Retry fields in the SOA RR are set too low, the secondary server will initiate zone transfers frequently. This error is called *zone thrash*; it results in more workload on both the primary and secondary name servers. Such incorrect data or increased workload may result in denial of service.

- **Threat T11—Information for Targeted Attacks:** RRs such as HINFO and TXT provide information about software name and versions (e.g., for resources such as Web servers and mail servers) that will enable the well-equipped attacker to exploit the known vulnerabilities in those software versions and launch attacks against those resources.

5.4   Security Objectives

Common objectives with respect to protection of the DNS host platform, DNS software, and DNS data are integrity and availability.

5.5   Host Platform Protection Approach

The protection/threat mitigation approaches for the DNS host platform consist of the following:

- Running a secure OS

- Secure configuration/deployment of OS.

These approaches are discussed in Section 7.1.

5.6   DNS Software Protection Approach

Best practice protection approaches for DNS software are as follows:

- Running the latest version of name server software, or an earlier version with appropriate patches

- Running name server software with restricted privileges

- Isolating name server software

- Setting up a dedicated name server instance for each function

- Removing name server software from nondesignated hosts

- Creating a topological and geographic dispersion of authoritative name servers for fault tolerance

- Limiting IT resource information exposure through two different zone files in the same physical name server (termed as split DNS) or through separate name servers for different client classes.

These approaches are described in Sections 7.2.1 through 7.2.9.

5.7   DNS Data Content Control – Protection Approach

Control of undesirable content in the zone file is accomplished by analyzing the contents for security implications, formulating integrity constraints that will check for the presence of such contents and verifying the zone file data for satisfaction of those constraints. Therefore, the only protection approach is to develop the zone file integrity checker software that contains the necessary constraints and can be run against the zone file to flag those contents that violate the constraints. To aid in formulation of constraints, desirable field values (ranges or lists) in the various RRs of zone file are required. These constraints need to be developed not only for RRs in an unsigned zone but also for additional RRs in a signed zone (zones that have implemented the DNSSEC specification). Hence, the recommendations for control of content of zone files are deferred to Section 10 after discussion of deployment guidelines for DNSSEC in Section 9 so that they cover those additional RRs as well.

The services provided by DNS also face threats resulting from vulnerabilities in network infrastructure components such as routers. Network configuration issues are outside the scope of this guidance document, however.

6. DNS Transactions—Threats, Security Objectives, and Protection Approaches

The threats to a DNS transaction depend on the type of transaction. Name resolution queries and responses (DNS query/response) between DNS clients (stub resolver or resolving name server) and DNS servers (caching/resolving name server or authoritative name server) could involve any nodes in the Internet; hence, the threats against them are much greater in number and severity compared to those for zone transfer, dynamic update, and DNS NOTIFY transactions. In general, the nodes involved in zone transfer, dynamic update, and DNS NOTIFY transactions are all within the administrative domain of a single organization. The only exceptions are instances in which the primary or secondary name servers of an organization are run on its behalf by ISPs or other organizations. There usually is a preexisting trust relationship in these cases, however, so it is not difficult to set up a mutual authentication system for DNS zone transfers.

This section seeks to identify protocol-based threats to the operation and administration of the DNS. This section also lists DNS protocol methods to address these threats. Non-DNS protocol based solutions such as IPsec are beyond the scope of this document but could be a more appropriate solution for an organizations based on its infrastructure.

6.1 DNS Query/Response Threats and Protection Approaches

DNS name resolution queries and responses (DNS query/response) generally involve single, unsigned, and unencrypted UDP packets. The known threats to DNS query/response transactions have been documented in IETF RFC 3833 [RFC3833] and can be classified as follows:

- **Threat T12:** Forged or bogus response

- **Threat T13:** Removal of some RRs from the response

- **Threat T14:** Incorrect expansion rules applied to wildcard RRs in a zone file.

6.1.1 Forged or Bogus Response

A forged or bogus response is a response that is different from the one that is expected from a legitimate authoritative name server. A bogus response can originate from:

- A compromised authoritative name server (for queries originating from a resolving name server)

- A poisoned cache of a resolving name server (for queries originating from a stub resolver).

An authoritative name server could be compromised by a platform-level attack on its OS or communication stack (see Section 5.1).

The cache of a resolving (caching) name server could be poisoned by the following attacks:

- **Packet Interception.** In this type of attack, the attacker eavesdrops on a request and is able to generate and send a response by spoofing an authoritative name server before the real response from the legitimate authoritative name server reaches the resolving name server.

- **ID Guessing and Query Prediction.** In this type of attack, the attacker guesses the ID field in the header of the DNS request message (because this field is only 16 bits long, brute force guessing is possible) and possibly the QNAME and QTYPE (owner name and RRType, respectively). The attacker then injects bogus data into the network as a response by spoofing a name server.

- **Responses Accumulated from a Compromised Authoritative Name Server**. A compromised authoritative name server is directed by a controlling adversary to send out bogus responses to queries from resolving name servers.

The impacts on a system serviced by a resolving name server that has a poisoned cache are as follows:

- **Denial of Service.** If some crucial RRs such as address records (A RRs) are forged, the system that requires this information can never establish connectivity with the intended node.

- **Client Redirection through Cache Poisoning.** Client redirection is performed by selective poisoning of DNS RRs whose RDATA element contains a name. Examples of such RRs are CNAME, NS, and MX. The name resolution (i.e., IP address) information for these names is found in a set of additional information (or *glue records* when discussing a delegation response). Normally the resolving name server obtains these necessary A/AAAA RRs through follow-up queries (also called *triggered queries*). The responses flowing into the network from these follow-up queries present yet another opportunity for the attacker to insert bogus records. First the attacker can introduce arbitrary names of the attacker's choosing in the RDATA portion of selected RRs; then the attacker can insert the IP addresses of servers (chosen by the attacker) in associated glue records that are transmitted as an answer to follow-up queries. This type of attack on two sets of related responses is called a *name chaining attack*. The overall effect of poisoning the cache of a resolving name server this way is to misdirect several clients who are making use of the services of that resolving name server. Redirecting the users to nodes of the attacker's choosing may enable the attacker to capture sensitive information such as passwords.

6.1.2   Removal of Some RRs

Apart from injecting bogus or forged data in a response, an attacker also could remove RRs from a response. This action might result in a name resolution query failure and consequent denial of service.

6.1.3   Incorrect Expansion Rules Applied to Wildcard RRs

Many zones use wildcard RRs to economize on the volume of data in the zone file. The wildcard patterns are used for synthesizing RRs on the fly in generating responses for name resolution queries. (The synthesis rules are outlined in section 4.3.2 of IETF RFC 1034 [RFC1034].) If synthesis rules are applied incorrectly in a name server, the RRs associated with resources existing in an organization may not be generated and made available in a DNS response. This fault also results in denial of service.

6.1.4   Protection Approach for DNS Query/Response Threats—DNSSEC

The underlying feature in the major threat associated with DNS query/response (i.e., forged response or response failure) is the integrity of DNS data returned in the response. Hence, the security objective is to verify the integrity of each response received. An integral part of integrity verification is to ensure that valid data has originated from the right source. Establishing trust in the source is called *data origin authentication*. Hence, the security objectives—and consequently the security services—that are required for securing the DNS query/response transaction are data origin authentication and data integrity verification.

These services could be provided by establishing trust in the source and verifying the signature of the data sent by that source. The specification for a digital signature mechanism in the context of the DNS infrastructure is in IETF's DNSSEC standard. The objectives, additional RRs, and DNS message contents involved in the DNSSEC are specified through RFCs 4033, 4034, and 4035 [RFC4033], [RFC4034], [RFC4035]. In DNSSEC, trust in the public key (for signature verification) of the source is established not by going to a third party or a chain of third parties (as in public key infrastructure [PKI] chaining), but by starting from a trusted zone (such as the root zone) and establishing the chain of trust down to the current source of response through successive verifications of signature of the public key of a child by its parent. The public key of the trusted zone is called the *trust anchor*.

After authenticating the source, the next process DNSSEC calls for is to authenticate the response. This requires that responses consist of not only the requested RRs but also an authenticator associated with them. In DNSSEC, this authenticator is the digital signature of an RRSet. The digital signature of an RRSet is encapsulated through a special RRType called RRSIG. The DNS client using the trusted public key of the source (whose trust has just been established) then verifies the digital signature to detect if the response is valid or bogus.

To ensure that RRs associated with a query are really missing in the zone file and have not been removed in transit, the DNSSEC mechanism provides a means for authenticating the nonexistence of an RR. It generates a special RR called an NSEC (or NSEC3) RR that lists the RRTypes associated with an owner name as well as the next name in the zone file. It sends this special RR, along with its signature, to the resolving name server. By verifying this signature, a DNSSEC-aware resolving name server can determine which authoritative owner name exists in a zone and which authoritative RRTypes exist at those owner names.

To protect against the threat of incorrect application of expansion rules for wildcard RRs, the DNSSEC mechanism provides a means of comparing the validated wildcard RR against an NSEC (or NSEC3) RR and thereby verifying that the name server applied the wildcard expansion rules correctly in generating an answer.

DNSSEC can guarantee the integrity of name resolution responses to DNS clients acting on behalf of Internet-based resources, provided the clients perform the DNSSEC signature verification. In many cases, however, these DNS clients are stub resolvers that are not DNSSEC-aware. If signature verification is performed by the resolving name server providing name resolution service for the clients that are stub resolvers, the end-to-end integrity of the response data can be guaranteed only by protecting the communication channel between the resolving name server and the stub resolver.

IETF's design criteria consider DNS data to be public; hence, confidentiality is not one of the security goals of DNSSEC. DNSSEC is not designed to directly protect against denial-of-service threats, although it does so indirectly by providing message integrity and source authentication. DNSSEC also does not provide communication channel security because name resolution queries and responses travel over millions of nodes of the public Internet. DNSSEC also can lead to a new type of weakness that did not exist in DNS before. An artifact of how DNSSEC performs negative responses allows a client to map all the names in a zone. This is called Zone Walking. Zone Walking provides an attacker with a "map" of a target zone with all domain names and IP addresses in the zone and enables him/her to determine the configuration of the internal network and launch some targeted attacks on some key hosts. Therefore, it is advisable that a zone only contains zone data that the administrator wants to be made public or use the

NSEC3 RR option for providing authenticated denial of existence. For internal DNS, something like split-DNS (see Section 7.2.8) could be deployed. For a discussion of NSEC3, see Section 10.4.

6.2 Zone Transfer Threats and Protection Approaches

Zone transfers are performed to replicate zone files in multiple servers to provide a degree of fault tolerance in the DNS service provided by an organization. Threats from zone transfers have not been documented formally through any IETF RFCs. A few threats could be expected, however: the first threat, denial of service, is common for any network transaction. The second threat is based on exploitation of knowledge gained from the information provided by zone transfers. The latter threat is common to any network packet.

- **Threat T15—Denial of Service:** Because zone transfers involve the transfer of entire zones, they place substantial demands on network resources relative to normal DNS queries. Errant or malicious frequent zone transfer requests on the name servers of the enterprise can overload the master zone server and result in denial of service to legitimate users.

- **Threat T16—**The zone transfer response message could be tampered.

The denial-of-service threat (T15) can be minimized if servers allowed to make zone transfer requests are restricted to a set of known entities. To configure this restriction into the primary name server, there should be a means of identifying those entities. Name server software such as BIND initially provided a configuration feature to restrict zone transfer requests to a set of designated IP addresses. Because IP addresses can be spoofed, however, this mode of configuration does not provide an adequate means of restricting zone transfer access.

The IETF developed an alternate mechanism called a *transaction signature* (TSIG), whereby mutual identification of servers is based on a shared secret key. Because the number of servers involved in zone transfer is limited (generally restricted to name servers in the same administrative domain of an organization), a bilateral trust model that is based on a shared secret key may be adequate for most enterprises (except for very large ones). TSIG specifies that the shared secret key be used not only for mutual authentication but also for signing zone transfer requests and responses. Hence, it provides protection against tampering of zone transfer response messages (threat T16). Protection of DNS data alone (the payload) in a zone transfer message also can be ensured through verification of signature records accompanying RRs from a DNSSEC-signed zone. These signatures, however, do not cover all the information in a zone file (e.g., delegation information). Furthermore, they enable verification of only the individual RRsets and not the entire zone transfer response message.

There is also another method to authenticate DNS transactions by using asymmetric cryptography (i.e. public key cryptography). The format of the SIG(0) RR is similar to the resource record signature (RRSIG) RR (see Section 9.2.1), and can be validated using a public key stored in the DNS (instead of a shared secret key). SIG(0) can be more computational expensive to use, but offer an advantage in that a previous trust relationship may not be necessary to use SIG(0) signed messages. However, since most zone transfers occur between parties that have a previously established relationship, it is considered easier to implement TSIG for authenticating zone transfer transactions.

Another possibility is to rely on lower level network layer to provide security such as IPSec. This would remove the need for authentication at the DNS (application) layer. How to set up this level of security is beyond the scope of this guide.

6.3 Dynamic Updates Threats and Protection Approaches

Dynamic updates involve DNS clients making changes to zone data in an authoritative name server in real time. Clients typically performing dynamic updates are CA servers, DHCP servers, or Internet Multicast Address servers. As with zone transfer transaction, the threats associated with dynamic update transaction have not been officially documented by the IETF through an RFC. The following are some common threats that could be expected, based on the fact that dynamic updates involve a data update request transiting a network.

- **Threat T17—Unauthorized Updates:** Unauthorized updates could have several harmful consequences for the content of zone data. Some harmful data operations include: (a) adding illegitimate resources (new FQDN and new RRs to a valid zone file), (b) deleting legitimate resources (entire FQDN or specific RRs), and (c) altering delegation information (NS RRs pointing to child zones).

- **Threat T18:** The data in a dynamic update request could be tampered.

- **Threat T19—Replay Attacks:** Update request messages could be captured and resubmitted later, thus causing inappropriate updates.

Threats T17 and T18 could be countered by authenticating the entities involved and providing a means to detect tampering of the messages. Because these security objectives in the case of zone transfer are met by the TSIG/SIG(0) mechanism, the same TSIG/SIG(0) mechanism is specified for protecting dynamic updates. Although the dynamic update message contains some replay attack (Threat T19) protection in the prerequisite field of the message, TSIG/SIG(0) provides an additional mechanism to protect against replay attacks by including a timestamp field in the dynamic update request. This signed timestamp enables a server to determine whether the timing of the dynamic update request is within the acceptable time limits specified in the configuration.

It sometimes makes more sense to use SIG(0) protection mechanisms for dynamic update than for zone transfer. Dynamic update transactions may happen between parties that do not always have a prior security relationship or may be part of a bootstrapping operation. Therefore it may be impractical to use TSIG with a shared secret, but SIG(0) authentication using keys stored in the DNS may be a possibility.

Another possibility is to rely on lower level network layer to provide security such as IPSec. This would remove the need for authentication at the DNS (application) layer. How to set up this level of security is beyond the scope of this guide.

6.4 DNS NOTIFY Threats and Protection Approaches

DNS NOTIFY is a message sent by primary (master) name servers to secondary (slave) name servers, causing the secondary servers to start a refresh operation (i.e. query for SOA RR to check the serial number, etc.) and perform a zone transfer if an update to the zone has occurred. Because the NOTIFY message is only a signal, there are only minor security risks in dealing with the message. The primary security risk to consider is the following:

- **Threat T20—Spurious NOTIFY Messages:** Secondary name servers would receive spurious DNS NOTIFY messages from sources other than the primary name server.

The only impact of receiving spurious DNS NOTIFY message is the increase in workload in secondary name servers since a zone transfer will only occur when an updated zone is on the primary server.

Because this threat is low impact, the only protection approach required is to configure the secondary name servers to receive DNS NOTIFY message only from the enterprise's primary name server. However, if TSIG is set up for use for all communication between a set of hosts, TSIG will be used with NOTIFY messages as well.

6.5 Threats Summary

Table 6-1 summarizes the DNS transactions and their associated threats, security objectives, and IETF security mechanism specifications to meet those goals.

Table 6-1. DNS Transaction Threats and Security Objectives

DNS Transaction	Threats	Security Objectives	DNS Based Security Specifications
DNS Query/Response	(a) Forged or bogus response (b) Removal of records (RRs) in responses (c) Incorrect application of wildcard expansion rules	(a) Data origin authentication (b) Data integrity verification	DNSSEC
Zone Transfer	(a) Denial of service (b) Tampering of messages	(a) Mutual authentication (b) Data integrity verification	TSIG
Dynamic Update	(a) Unauthorized Updates (b) Tampering of messages (c) Replay attack	(a) Mutual authentication (b) Data integrity verification (c) Signed timestamps	TSIG, GSS-TSIG or SIG(0)
DNS NOTIFY	(a) Spurious notifications	(a) To prevent denial of service through increase in workload	Specify hosts from which this message can be received TSIG or SIG(0)

7. Guidelines for Securing DNS Hosting Environment

The guidelines for secure configuration of the DNS hosting environment are classified under the following headings:

- Securing DNS host platform

- Securing DNS software

- Content control of zone file.

These guidelines are provided mostly in the context of BIND DNS Name Server software. Where possible, similar guidelines will be provided for other DNS authoritative software packages such as NSD (from NLnet Labs) and Microsoft Windows Server.

7.1 Securing DNS Host Platform

The platform on which the name server software is hosted should be running an adequately secured OS. Most of the DNS installations run either on a flavor of Unix or Windows. Given this scenario, it is necessary to ensure the following:

- The latest OS patches are installed.

- Recommended OS configuration practices as issued by CERT®//CC [CERT] and NIST's NVD metabase [NVD], based on identified vulnerabilities that pertain to the application profile into which the name server software fits, are followed. In particular, hosts that run the name server software should not provide any other services and therefore should be configured to respond to DNS traffic only. In other words, the only allowed incoming ports/protocols to these hosts should be 53/udp and 53/tcp. Outgoing DNS messages should be sent from a random port to minimize the risk of an attacker guessing the outgoing message port and sending forged replies.

7.2 Securing DNS Software

Protection approaches for DNS software include choice of appropriate version, installation of patches, running it with restricted privileges, restricting other applications in the execution environment, dedicating instances for each function, controlling the set of hosts where software is installed, placement within the network, and limiting information exposure by logical/physical partitioning of zone file data or running two name server software instances for different client classes.

7.2.1 Running the Latest Version of Name Server Software

Each newer version of the name server software, especially the BIND software, generally is devoid of vulnerabilities found in earlier versions because it has design changes incorporated to take care of those vulnerabilities. Of course, these vulnerabilities have been exploited (i.e., some form of attack was launched), and sufficient information has been generated with respect to the nature of those exploits. Thus, it makes good business sense to run the latest version of name server software because theoretically it is the safest version. Even if the software is the latest version, it is not safe to run it in default mode. The security administrator should always configure the software to run in the recommended secure mode of operation after becoming familiar with the new security settings for the latest version.

In some installations, it may not be possible to switch over to the latest version of name server software immediately. In these situations, the administrator should keep pace with vulnerabilities identified in the operational version and associated security patches [BINDSEC].

Checklist item 1: When installing the upgraded version of name server software, the administrator should make necessary changes to configuration parameters to take advantage of new security features.

Checklist item 2: Whether running the latest version or an earlier version, the administrator should be aware of the vulnerabilities, exploits, security fixes, and patches for the version that is in operation in the enterprise. The following actions are recommended (for BIND deployments):

- Subscribe to ISC's mailing list called "bind-announce" or "nsd-users" for NSD

- Periodically refer to the BIND vulnerabilities page at
  http://www.isc.org/products/BIND/bind-security.html

- Refer to CERT®/CC's Vulnerability Notes Database at http://www.kb.cert.org/vuls/ and the NIST NVD metabase at http://nvd.nist.gov/.

7.2.2   Turning off the Version Query

There is a feature in BIND that returns the version number of the server daemon running if a special query is sent to the server. This query is for the string "version.bind" with query type TXT and query class Chaos (CH). This information may be of use to attackers who are looking for a specific version of BIND with a discovered weakness. BIND can be configured to refuse this type of query by having the following command in the BIND configuration file (`/etc/named.conf`).

```
options {
      version none;
};
```

There is a similar feature in NSD to refuse version queries. In the configuration file for NSD (`/etc/nsd/nsd.conf`).

```
server:
      hide-version: yes
```

For Windows, the DNS version number can be changed using the **dnscmd** tool with the `/config` option and changing the `EnableVersionQuery` property[3].

Checklist item 3: To prevent information about which version of name server software is running on a system, name servers should be configured to refuse queries for its version information..

---

[3] http://msdn.microsoft.com/en-us/library/cc422472(v=prot.10).aspx

7.2.3 Running Name Server Software with Restricted Privileges

If the name server software is run as a privileged user (e.g., root in Unix systems), any break-in into the software can have disastrous consequences in terms of resources resident in the name server platform. Specifically, a hacker who breaks into the software acquires unrestricted access and therefore can potentially execute any commands or modify or delete any files. Hence, it is necessary to run the name server software as a nonprivileged user with access restricted to specified directories to contain damages resulting from break-in.

The user name under which the name server software needs to run can be specified by using the chroot command in BIND. This is why this approach is known as running the DNS server in a chroot jail. An example command (entered on the command line) is the following:

```
>named -u named -t /var/named/chroot
```

where the options specify the following:

–u specifies the userID to which the name server software will change after starting. This user account should be created prior to issuing the chroot command.

–t specifies the directory in which the name server software owner will use as the root "jail" and should have the appropriate privileges. This directory will have to be created by the system administrator before starting the server process.

This ability is also found in NSD as part of the NSD startup scripts. The default userID under which NSD runs is nsd and the default directory is /etc/nsd/ Both of these defaults can be changed in the NSD configuration file by adding:

```
server:
      username: <userID>
      chroot: <dir>
```

7.2.4 Isolating the Name Server Software

Even if the DNS software (e.g., BIND) is run on a secure OS, the vulnerabilities of other software programs on that platform can breach the security of DNS software. Hence, it is recommended that the platform on which the DNS software runs contains no programs other than those needed for OS and network support. If this is not possible due to resource constraints, care should be taken to ensure to limit the number of services running on the same platform.

7.2.5 Dedicated Name Server Instance for Each Function

An authoritative name server serves RRs from its own zone file; this function is called an *authoritative function*. Serving RRs either from its cache (directly or by building up its cache dynamically through iterative queries) is called a *resolving function*; this is how a resolving name server provides responses. A name server instance can be configured as an authoritative name server, a resolving name server, or both. Because of attacks such as cache poisoning (see Section 6.1), however, a resolving name server has to be run under security policy that is different from that of an authoritative name server. Hence, a name server instance should always be configured as either an authoritative name server or a resolving name server.

An authoritative name server is only intended to provide name resolution for the zones for which it has authoritative information. Hence, the security policy should have recursion turned off for this type of name server. Disabling recursion prevents an authoritative name server from sending queries on behalf of other name servers and building up a cache using responses. Disabling this function eliminates the cache poisoning threat on authoritative serves and prevents their use as reflectors for DDoS attacks [BCP140]. In BIND, recursion is disabled by using the options statement in the BIND configuration file as follows:

```
options {
       recursion no;
};
```

NSD can only operate as an authoritative only server. NSD cannot act as a recursive resolver and therefore does not have a comparable option, as it is NSD's default behavior.

A resolving name server is only intended to provide resolving services (processing resolving queries on behalf of clients) for internal clients. Thus, protection of resolving name servers can be ensured by restricting their types of interactions (also called transactions) to designated hosts through various configuration options in the configuration file of the name server software.

7.2.6   Removing Name Server Software from Nondesignated Hosts

DNS software should not be running or present in hosts that are not designated as name servers. The possibility arises in the case of DNS BIND software because of the fact that many versions of Unix (including Solaris and Linux versions) come installed with BIND as default. Hence, while taking an inventory of software in workstations and servers of the enterprise as part of the security audit, it is necessary to look for BIND installations and remove them from hosts that are not functioning as name servers.

7.2.7   Network and Geographic Dispersion of Authoritative Name Servers

Most enterprises have an authoritative primary server and a host of authoritative secondary name servers. It is essential that these authoritative name servers for an enterprise be located on different network segments. This dispersion ensures the availability of an authoritative name server not only in situations in which a particular router or switch fails but also during events involving an attack on an entire network segment. In addition to network-based dispersion, authoritative name servers should be dispersed geographically as well. In other words, in addition to being located on different network segments, the authoritative name servers should not all be located within the same building. One approach that some organizations follow is to locate some authoritative name servers in their own premises and others in their ISPs' data centers or in partnering organizations.

Additionally, a network administrator may choose to use a "hidden" master authoritative server and only have secondary servers visible on the network. A hidden master authoritative server is an authoritative DNS server whose IP address does not appear in the name server set for a zone. All of the name servers that do appear in the zone database as designated name servers all get their zone data from the hidden master via a zone transfer request. In effect, all visible name servers are actually secondary slave servers. This prevents potential attackers from targeting the master name server, as its IP address may not appear in the zone database. A hidden master only accepts zone transfer requests from the set of valid secondaries and refuses all other DNS queries.

Checklist item 4: The authoritative name servers for an enterprise should be both network and geographically dispersed. Network-based dispersion consists of ensuring that all name servers are not behind a single router or switch, in a single subnet, or using a single leased line. Geographic dispersion consists of ensuring that not all name servers are in the same physical location, and hosting at least a single secondary server off-site.

Checklist item 5: If a hidden master is used, the hidden authoritative master server should only accept zone transfer requests from the set of secondary zone name servers and refuse all other DNS queries. The IP address of the hidden master should not appear in the name server set in the zone database.

7.2.8   Limiting Information Exposure through Partitioning of Zone files

Authoritative name servers for an enterprise receive requests from both external and internal clients. In many instances, external clients need to receive RRs that pertain only to public services (public Web server, mail server, etc.) Internal clients need to receive RRs pertaining to public services as well as internal hosts. Hence, the zone information that serves these RRs can be split into different physical files for these two types of clients: one for external clients and one for internal clients. This type of implementation of the zone file is called *split DNS*.

Split DNS does have some drawbacks. First, remote hosts (travelers using laptops to connect back to an organization, for example), may not be using an internal resolving DNS server, and therefore may not be able to see internal hosts. Second, internal host information may leak to outside the firewall (by accident or attack), defeating the purpose of having a split DNS, or causing confusion of an internal and external host have the same FQDN, but different IP addresses. Split DNS should not be seen as a replacement for proper access control techniques.

Checklist item 6: For split DNS implementation, there should be a minimum of two physical files or *views*. One should exclusively provide name resolution for hosts located inside the firewall. It also can contain RRsets for hosts outside the firewall. The other file or view should provide name resolution only for hosts located outside the firewall or in the DMZ, and not for any hosts inside the firewall.

To set up split DNS using BIND, the view statement is used in the BIND configuration file. For example, to set up an authoritative view of the zone "sales.mycom.com" for internal clients:

```
view "insider" {
      match-clients { internal_hosts; };
      recursion no;
      zone sales.mycom.com {
            type master;
            file "sales_internal.db";
      };
};
```

In this statement, the file "sales_internal.db" contains the authoritative information for zone "sales.mycom.com" and restricts that information to queries coming from the address list named "internal_hosts". Section 8.1.1 provides details on how to set up this list. To set up a view of the same zone, but with only external hosts for queries coming from outside the network, the following view statement is used in the same configuration file:

```
view "outsider" {
      match-clients { any; };
      match-destinations { public_hosts; };
      recursion no;
      zone sales.mycom.com {
            type master;
            file "sales_external.db";
      };
};
```

This statement is very similar to the previous statement, except that the file with the zone view is "sales.external.db" and the client list is given as "any", meaning that queries coming from both outside and inside the network can see the same external view of "sales.mycom.com." Together, these statements allow internal clients to view both the internal and external hosts in "sales.mycom.com," whereas external hosts (not on the same network) can see only DNS information contained in the external view of the zone where the destination matches the address list "public_hosts."

7.2.9 Limiting Information Exposure Through Separate Name Servers for Different Clients

Instead of having the same set of authoritative name servers serve different types of clients, an enterprise could have two different sets of authoritative name servers. One set, called *external name servers*, can be located within a DMZ; these would be the only name servers that are accessible to external clients and would serve RRs pertaining to hosts with public services (Web servers that serve external Web pages or provide B2C services, mail servers, etc.) The other set, called *internal name servers*, is to be located within the firewall and should be configured so they are not reachable from outside and hence provide naming services exclusively to internal clients. The purpose of both architecture options (i.e., two different sets of name servers and split DNS) is to prevent the outside world from knowing the IP addresses of internal hosts. This configuration may be the only available option for enterprises that use DNS server software that does not have the view feature found in BIND or organizations that use NSD as their authoritative DNS server.

7.3 Content Control of Zone File

As stated in Section 5.7, the only protection approach for content control of DNS zone file is the use of a zone file integrity checker. The effectiveness of integrity checking using a zone file integrity checker depends upon the database of constraints built into the checker. Hence, the deployment process consists of developing these constraints with the right logic and the only determinant of the truth value of these logical predicates are the parameter values for certain key fields in the format of various RRTypes. The choice of these parameter values forms the deployment guidelines and is discussed in Section 10.

7.4 Recommendations Summary

The following items provide a summary of the major recommendations from this section:

- **Checklist item 1:** When installing the upgraded version of name server software, the administrator should make necessary changes to configuration parameters to take advantage of new security features.

- **Checklist item 2:** Whether running the latest version or an earlier version, the administrator should be aware of the vulnerabilities, exploits, security fixes, and patches for the version that is in operation in the enterprise. The following actions are recommended:

    – Subscribe to ISC's mailing list called "bind-announce" or NLnet Labs mailing list "nsd-users"

    – Periodically refer to the BIND vulnerabilities page at http://www.isc.org/products/BIND/bind-security.html

    – Refer to CERT®/CC's Vulnerability Notes Database at http://www.kb.cert.org/vuls/ and the NIST NVD metabase at http://nvd.nist.gov/.

    For other implementations (e.g., MS Windows Server), other announcement lists may exist.

- **Checklist item 3:** To prevent information about which version of server software is running on a system, name servers should be configured to refuse queries for its version.

- **Checklist item 4:** The authoritative name servers for an enterprise should be both network and geographically dispersed. Network-based dispersion consists of ensuring that all name servers are not behind a single router or switch, in a single subnet, or using a single leased line. Geographic dispersion consists of ensuring that not all name servers are in the same physical location, and hosting at least a single secondary server off-site.

- **Checklist item 5:** If a hidden master is used, the hidden authoritative master server should only accept zone transfer requests from the set of secondary zone name servers and refuse all other DNS queries. The IP address of the hidden master should not appear in the name server set in the zone database.

- **Checklist item 6:** For split DNS implementation, there should be a minimum of two physical files or views. One should exclusively provide name resolution for hosts located inside the firewall. It also can contain RRsets for hosts outside the firewall. The other file or view should provide name resolution only for hosts located outside the firewall or in the DMZ, and not for any hosts inside the firewall.

8. Guidelines for Securing DNS Transactions

Section 6 outlines threats, security objectives, and protection approaches for various DNS transactions. This section provides the steps involved in implementing those approaches, as well as operational best practices that go with those implementations. The DNS protocol protection approaches in Table 6-1 are aggregated, indexed by type of approach, and elaborated here as follows:

- **Restricting Transaction Entities Based on IP Address.** In this type of implementation, the DNS name servers and clients participating in a DNS transaction are restricted to a trusted set of hosts by specifying their IP addresses in appropriate access control statements provided by the name server software. The protection provided by these IP-based access control statements can be circumvented by attacks such as IP spoofing. Hence, this solution is not recommended as the sole protection mechanism for DNS query/response, zone transfer, and dynamic update transactions that have high threat impact. However, for the DNS NOTIFY transaction, where the only threat is spurious notification (which may not even trigger a zone transfer), an access control based on IP address will suffice. Although this solution is not recommended generally, a description of the mechanics of access control using IP addresses is provided in Section 8.1 because the same statements are used to identify hosts based on named keys while implementing transaction protection using hash-based message authentication codes. This approach has been implemented for all DNS transactions.

- **Transaction Protection through Hash-Based Message Authentication Codes (TSIG Specification).** In this approach, transaction protection is enabled through generation and verification of hash-based message authentication codes (HMAC). Because these codes are embedded within a special RR of RRType TSIG, the specifications that outline protection of DNS transactions using HMAC are called TSIG in the DNS community. TSIG specifications are described in RFC 2845 and 3007 [RFC2845], [RFC3007]. Application of TSIG specifications for protection of zone transfer and dynamic update transactions is described in Section 8.2.

- **Transaction Protection through Asymmetric Digital Signatures (DNSSEC Specification).** This approach, which goes by the name DNS security extensions (DNSSEC), is described through a family of RFCs [RFC4033], [RFC4034], and [RFC4035]. The core services provided by DNSSEC are data origin authentication and integrity protection. DNSSEC is used mainly for securing DNS information obtained from DNS query/response transactions. The deployment issues in DNSSEC are described in Section 9.

8.1 Restricting Transaction Entities Based on IP Address

Some DNS name server implementations, such as BIND 9.x, provide access control statements through which it is possible to specify hosts that can participate in a given DNS transaction. The hosts can be identified by their IP address or IP subnet reference (called *IP prefix*) in these statements.

The list containing these IP addresses and/or IP prefixes is called an *address match list*. (An address match list can be made up of other things besides IP addresses and IP prefixes, as described in Section 8.1.1). The address match list is used as an argument in various access control statements that are available for use in BIND configuration files. There are separate access control statements for each type of DNS transaction. The syntax of the various access control statements and the DNS transaction for which each is used are given in Table 8.1.

Table 8-1. BIND Access Control Statement Syntax for DNS Transactions

Access Control Statement Syntax	DNS Transaction
allow-query { address_match_list }	DNS Query/Response
allow-recursion { address_match_list }	Recursive Query
allow-transfer { address_match_list }	Zone Transfer
allow-update { address_match_list }	Dynamic Update
allow-update-forwarding { address_match_list }	Dynamic Update
allow-notify { address_match_list }	DNS NOTIFY
blackhole { address_match_list }	Blacklisted Hosts

The purpose of each of these access control statements is as follows:

- **allow-query:** specifies the list of hosts allowed to query the name server as a whole or a particular zone within the name server

- **allow-recursion:** specifies the list of hosts allowed to submit recursive queries to the name server as a whole or to a particular zone served by the name server

- **allow-transfer:** specifies the list of hosts allowed to initiate zone transfer requests to the name server as a whole or to a particular zone within the name server. This statement is predominantly required for configuration of master name servers.

- **allow-update:** specifies the list of hosts allowed to initiate dynamic update requests

- **allow-update-forwarding:** specifies the list of hosts allowed to forward dynamic update requests (regardless of the originator of the requests)

- **allow-notify:** specifies the list of hosts from which to accept DNS NOTIFY messages indicating changes in the zone file. This list is relevant only for configuration of secondary slave name servers.

- **blackhole:** specifies the list of hosts that are blacklisted (barred) from initiating any transaction with this name server. Used only in an "options" server wide ACL statement.

The foregoing access control statements are, in fact, substatements that can be used in the context of *options* and *zone* statements in the BIND 9.x configuration file (with the exception of **blackhole**). When they are used within the zone statement, they specify access control restrictions for the corresponding DNS transaction for that specific zone. When they are used as part of the options statement, they specify access control restrictions for the corresponding DNS transaction for the name server as a whole (because a name server could host multiple zones).

NSD has a similar set of configuration options for certain transactions. NSD has a more limited sets of options and currently only restrict zone transfers. In the following sections, if a comparable set of options exists for NSD, they will be listed. Otherwise it should be noted that a comparable option does not exist at the time of writing.

8.1.1 Restricting DNS Query/Response Transaction Entities

An example of the usage of the allow-query substatement (to specify restrictions for the DNS query/response transaction stating the IP addresses/subnets from which DNS queries are accepted) both at the server level and at the zone level (for the zone example.com) is given below:

```
options {
      allow-query { 254.10.20.10; 239.10.30.29/25; };
};

zone "example.com." {
      type master;
      file "zonedb.example.com";
      allow-query { 192.249.249.1; 192.249.249.4; };
};
```

Specifying the list of IP addresses and IP prefixes within the options and zone statements could clutter the configuration file. Furthermore, the list of IP addresses and IP prefixes could be the same for many of the access control statements within a name server, and errors could be introduced if any additions or subtractions are made for that list. To avoid these problems, BIND provides a means to create named address match lists, which are called *access control lists* (ACL). These ACLs can be used in place of the list of IP addresses/IP prefixes (in the address match list argument) in the access control statements.

The ACLs are created by using the acl statement in BIND 9.x. The general syntax of the acl statement is as follows:

```
acl acl-list-name {
      address_match_list
};
```

The acl-list-name is a user-defined string (e.g., internal_hosts). The address_match_list can be a list of IP addresses, IP address prefixes (denoting subnets), or cryptographic keys. An example of an acl statement that uses an IP address and a subnet reference in address_match_list is given below. In the example, 254.10.20.10 denotes the IP address of a host, and the IP prefix 239.10.30.0/24 denotes a class C subnet.

```
acl "internal_hosts" {
      254.10.20.10;
      239.10.30.0/24;
};
```

The use of ACL – "internal_hosts" in place of the list of IP addresses/IP prefix in the options and zone statement given above is as follows:

```
options {
      allow-query { internal_hosts; };
};

zone "example.com." {
      type master;
      file "zonedb.example.com";
      allow-query { internal_hosts; };
};
```

The address match list parameter in an access control statement can contain any of the following values:

- An IP address or list of IP addresses

- An IP prefix or list of IP prefixes

- ACLs

- A combination of the above three.

The definition of ACLs forms a critical element in the configuration of DNS transaction restrictions. Hence, it is a good operational practice for the DNS administrator to define and create ACLs pertaining to different DNS transactions.

Checklist item 7: It is recommended that the administrator create a named list of trusted hosts (or blacklisted hosts) for each of the different types of DNS transactions. In general, the role of the following categories of hosts should be considered for inclusion in the appropriate ACL:

- DMZ hosts defined in any of the zones in the enterprise

- All secondary name servers allowed to initiate zone transfers

- Internal hosts allowed to perform recursive queries.

In addition to IP address, IP prefix, or ACL, the address match list parameter in the access control statements can take on any of the following special values:

- **none:** matches no hosts

- **any:** matches all hosts

- **localhost:** matches all IP addresses of the server on which the name server is running

- **localnets:** matches all IP addresses and subnet masks of the server on which the name server is running.

Following are a few more examples of commands for creating ACLs and the use of ACLs within options and zone statements:

```
acl "local_hosts" {
     254.10.20.10;
     239.10.30.29/25;
};

acl "fake-net" {
     0.0.0.0/8;
     1.0.0.0/8;
};
```

```
options {
     allow-query { any; };
     blackhole { fake-net; };
};

zone "example.com." {
     type master;
     file "zonedb.example.com";
     allow-query { local_hosts; };
};
```

In the named.conf snippet above, two ACLs, local_hosts and fake-net, have been specified. DNS queries from any hosts are allowed at the server level. No transactions are permitted from the hosts included under fake-net. Queries to the zone example.com can be initiated only by the hosts included under the ACL local_hosts because any restriction specified under the zone (zone-specific) statement overrides the restriction specified under the options (server-wide) statement.

Key material can also be used in ACL statements. This would indicate that only hosts knowing (and using) the shared key (or key pair) would be able to communicate. How a key is use in an ACL is discussed in Section 8.2.2.

NSD does not have the feature to define ACLs as a means of only allowing queries from designated set of hosts.

Administrators using Microsoft Windows Server as their DNS server can query the Microsoft TechNet[4] website for information on configuring access control lists for their server version.

8.1.1.1 Restricting Recursive Queries (a special case under DNS Query/Response)

Authoritative name servers provide name resolution service from their own data and are supposed to provide this service for any DNS client. Hence, configuring an authoritative name server to accept queries from a restricted set of hosts does not make sense. The practical security protection for an authoritative name server is to turn off the query recursion feature so that the authoritative name server does not poison its cache by querying other (possibly compromised) name servers. A local resolving/recursive name server can be configured to accept queries only from internal hosts, to protect it from denial-of-service attacks as well as cache poisoning. However, there may be situations in which it is economically infeasible to dedicate separate servers for authoritative service and resolution service, and the resolving name server has to perform as authoritative server for one or more zones. In this situation, the following strategies are possible within the BIND 9.x name server:

- Restricting all queries accepted by the server to a specified set of IP addresses of internal clients and then overriding this set only for authoritative zones so that any DNS client can obtain information for resources in that zone.

- Restricting recursive queries to a specified set of IP addresses of internal clients through a direct configuration option

- Serving different responses (data) to different clients by defining views.

---

[4] http://technet.microsoft.com/en-US/

<u>Restriction at server level with override for authoritative zones:</u>

In this strategy, the allowable set of internal clients who can submit queries to the name server is specified through the acl statement as follows:

```
acl internal_hosts {192.158.43.3; 192.158.43.6; 192.158.44.56;};
```

The server-wide option would be to restrict all queries to the list of clients:

```
options {
        allow-query { internal_hosts; };
        - or –
        allow-recursion { internal_hosts; };
        };
```

The option can be overridden by specifying zones for which this name server is authoritative (thus allowing queries to that zone from all clients):

```
zone "example.com" {
        type master;
        file "zonedb.example.com";
        allow-query { any; };
};
```

<u>Restricting all recursive queries to a specified set of IP addresses:</u>

Server-wide restriction:

```
options {
        allow-recursion { internal_hosts; };
};
```

<u>Restricting recursion through views:</u>

The purpose of creating views is to create a logical partition made up of a combination of clients (based on IP addresses) and zones for which recursive queries will be supported and those for which they will not be supported. In the following example, the view recursion_view is enabled to define the scope of IP addresses and zones that are permitted to submit recursive queries; no_recursion_view is meant for disallowing recursion.

```
view recursion_view {
      match-clients { internal_hosts; };
      recursion yes;
};

view no_recursion_view {
      match-clients { any; };
      recursion no;
};
```

It should be noted that NSD (as of the time of writing) is an authoritative only DNS server. Therefore, an NSD server will never act as a recursive server and only serve authoritative information from the zones it is configured to serve.

8.1.2   Restricting Zone Transfer Transaction Entities

Authoritative name servers (especially primary name servers) should be configured with an allow-transfer access control substatement designating the list of hosts from which zone transfer requests can be accepted. These restrictions address the denial-of-service threat and potential exploits from unrestricted dissemination of information about internal resources. Based on the need to know, the only name servers that need to refresh their zone files periodically are the secondary name servers. Hence, zone transfer from primary name servers should be restricted to secondary name servers. The zone transfer should be completely disabled in the secondary name servers. The address match list argument for the allow-transfer substatement should consist of IP addresses of secondary name servers and stealth secondary name servers.

The command to create an ACL "valid_secondary_NS" with the IP addresses of three secondary name servers is as follows:

```
acl "valid_secondary_NS" {
     224.10.229.5;
     224.10.235.6;
     239.10.245.25;
};
```

The allow-transfer substatement can be used in a zone statement and in an options statement. When it is used in a zone statement, it can restrict zone transfer for that zone; when it is used in an options statement, it can restrict zone transfer for all zones in the name server.

The allow-transfer substatement at the server level is as follows:

```
options {
     allow-transfer { "valid_secondary_NS"; };
};
```

The allow-transfer substatement at the zone level is as follows:

```
zone "example.com" {
     type master;
     file "zonedb.example.com";
     allow-transfer { "valid_secondary_NS"; };
};
```

The foregoing statements apply to primary name servers. In the secondary and stealth secondary name servers, zone transfer should be disabled as shown below:

```
zone "example.com" {
     type slave;
     masters { 224.239.5.1; } ;
     file "zonedb_bak.example.com";
     allow-transfer { none; };
};
```

8.1.3 Restricting Zone Transfer in NSD

NSD has a similar set of tools to restrict zone transfers to only a chosen set of slave servers. Like BIND, the administrator should learn and use the options available in the NSD configuration file. There is no way to create access control lists (ACLs), but an administrator can list the individual IP address of slave servers in the zone statements in the NSD configuration file.

In the configuration file, the provide-xfr statement is used in the zone statement block of the nsd.conf file much like a combined masters statement and allow-transfer statement in BIND configuration files:

```
zone:
        #allow transfer from subnet
        provide-xfr: 169.192.85.0/24
        #prevent transfer from specific IP address in block
        provide-xfr: 169.192.85.66 BLOCKED
```

Only one IP address should appear in a provide-xfr statement, but the address can be an entire subnet. The provide-xfr statement allows transfers; all other transfer requests are rejected by default.

8.1.4 Restricting Zone Transfer in Windows Server

Microsoft Windows DNS Server can also restrict zone transfers based on different criteria for increased operational security. Administrators can deny all zone transfer requests, or restrict zone transfers to a set of server (the servers in the NS set for example). Microsoft's TechNet Library[5] contains the current information on how to use the administrator tools to restrict zone transfers.

8.1.5 Restricting Dynamic Update Transaction Entities

Dynamic updates on a zone file can be directed only to the copy of the zone file that resides on the primary name server for the zone (i.e., where the master zone file resides). By default, dynamic update is turned off in both BIND 8 and BIND 9. Dynamic updates are enabled or restricted by using one of the following two statements in BIND:

- allow-update

- update-policy (available only in BIND 9 versions).

These statements can be specified only at the zone level, not at the server level. Hence, these statements are substatements within the zone statement. The allow-update substatement enables specification of dynamic update restrictions based on IP addresses and a shared secret (also called a *TSIG key*[6]). The use of the allow-update statement using IP addresses alone is addressed in this section. The use of the allow-update statement using TSIG keys is described in Section 8.2.6.

The update-policy statement enables specification of dynamic update restrictions based on TSIG keys only, but it enables specification of update restrictions at a finer level of granularity. The allow-update substatement implies update access rights to all records of a zone; the update-policy substatement can be used to restrict update access rights to one or more designated RRTypes (e.g., A RRs).

---

[5] http://technet.microsoft.com/en-us/library/cc771652.aspx
[6] The term TSIG Key (while commonly used) is not technically correct, as it refers to a shared secret string and not a cryptographic key.

To use the allow-update statement, an address match list must be created. The command to create an ACL DU_Allowed_List with one IP address is as follows:

```
acl "DU_Allowed_List" {
      192.249.12.21;
};
```

The ACL DU_Allowed_List (consisting of IP addresses of hosts allowed to send dynamic update requests for updating the contents of the zone example.com) is used within the allow-update substatement of the zone statement as follows:

```
zone "example.com" {
      type master;
      file "zonedb.example.com";
      allow-update { "DU_Allowed_List"; };
};
```

Dynamic update requests generally originate from hosts such as DHCP servers that assign IP addresses dynamically to hosts. Once they assign an IP address to a new host, they need to store the FQDN-to-IP address mapping (by creating an A RR) and address-to-FQDN mapping (by creating a PTR RR) information in the primary authoritative name servers for the zones. Creation of this information occurs through dynamic updates.

As of the time of writing, NSD does not support dynamic update so there are no comparable configuration options for NSD. All dynamic update messages sent to a DNS sever running NSD will be rejected. Updates to a zone must be done offline and then the server signaled to reload the new modified zone.

For details on how to secure dynamic update on Windows DNS servers, administrators should consult the Microsoft TechNet for details[7].

8.1.6   Restricting BIND DNS NOTIFY Transaction Entities

Once zone transfers have been set up between servers, it is a good idea to make sure that secondary name servers are informed about changes to zone file data through a notification message. By default, a notification message is sent whenever a primary name server detects a change in the zone file. It sends a DNS NOTIFY message to every name server listed in the NS RRSet in the zone because they are the recognized secondary name servers of the zone. DNS administrators should keep notification on because this configuration will allow updates to be propagated quickly to secondary name servers. If the DNS administrator wants to turn off the functionality for a specific zone, however, the notify substatement should be used in the zone statement of that zone:

```
zone "example.com" {
      type master;
      notify no;
      file "zonedb.example.com";
};
```

If there are any additional servers to which the zone administrator wants the DNS NOTIFY message to be sent (e.g., a stealth slave server), the "also-notify" substatement should be added to the zone statement, and the IP addresses of the additional servers should be specified as its parameter values, as shown below:

---

[7] http://technet microsoft.com/en-us/library/cc771255.aspx

```
zone "example.com" {
      type master;
      also-notify { 192.168.25.2; };
      file "zonedb.example.com";
};
```

The receiver of the DNS NOTIFY message, the secondary name server, allows notify messages only from the primary name server by default. (Recall that the secondary name server is made aware of its primary name server through the master substatement in the zone statement.) If the secondary name server wants to receive notify messages from additional servers, the "allow-notify" substatement in the zone statement must be added, and then the IP addresses of those servers must be specified in that substatement, as follows:

```
zone "example.com" {
      type slave;
      allow-notify { 193.168.25.4; };
      file "zonebak.example.com";
      masters { 192.168.25.1; };
};
```

8.1.7   Restricting NSD DNS NOTIFY Transaction Entities

There are two statements (both placed in the zone: statement block) that a DNS administrator can use to send DNS NOTIFY messages or restrict listening for DNS NOTIFY messages to a particular IP address (a master server in the case of NSD acting as a slave server).

To configure NSD to send DNS NOTIFY messages to a particular IP address (either a slave secondary or a stealth secondary) and a particular TSIG key or the option "NOKEY" if no TSIG is used, the following is added to the zone: statement block in the NSD configuration file:

```
zone:
         notify: 10.0.0.10   NOKEY
```

For a slave server the configuration statement in the zone: block to indicate which IP address to accept DNS NOTIFY messages from:

```
zone:
         allow-notify: 10.11.12.13   NOKEY
```

8.2   Transaction Protection Through Hash-Based Message Authentication Codes (TSIG)

The process of authenticating the source of a message and its integrity through hash-based message authentication codes (HMAC) is specified through a set of DNS specifications known collectively as TSIG. The term HMAC is used to denote both the message authentication code generated by using a keyed hash function and the hash function itself. HMAC is specified in RFC 2104 [RFC2104] and generalized in the NIST document FIPS 198-1 [FIPS198].

An HMAC function uses two parameters—a message input and a secret key—and produces an output called a *message authentication code* (MAC) or *hash*. The sender of the message uses the HMAC function to generate a MAC and sends this MAC along with the message to the receiver. The receiver,

SECURE DOMAIN NAME SYSTEM (DNS) DEPLOYMENT GUIDE

who shares the same secret key, uses the key and HMAC function used by the sender to compute the MAC on the received message. The receiver then compares the computed MAC with the received MAC; if the two values match, it provides assurance that the message has been received correctly and that the sender belongs to the community of users sharing the same secret key. Thus, message source authentication and integrity verification are performed in a single process.

The hash algorithm (which forms the primitive for the hash function) generates a fixed-size MAC or hash from a message of arbitrary size. The HMAC function for TSIG specified in RFC 2845 [RFC2845] and extended in RFC 4635 [RFC4635] to support more hash algorithms (SHA-1 and SHA-2 family of algorithms).

Transaction protection through HMAC using a shared secret is not a scalable solution. This is the reason the TSIG specification is largely used only for zone transfer and dynamic update transactions. These DNS transactions are either between servers in the same administrative domain or between servers in domains with previous established interactions.

The MAC or hash value generated by the sender of the DNS message is placed in a new RR called a TSIG record that is added to the DNS message. The TSIG record, in addition to the generated hash, contains the following:

- Name of hash algorithm

- Key name

- Time the hash was generated (timestamp)

- "Fudge factor" –time in seconds (usually 5 minutes) to use as delta on either side of the time generated for which the TSIG signature should be considered valid; used to account for possible clock skew between hosts.

The timestamp field specifies the time at which the MAC was generated. The purpose of this field is to protect against replay attacks. In a replay attack, the attacker could capture the packet containing the MAC and send it after a period of time. To ensure that this does not happen, the recipient reads the MAC generation time and the current clock time and verifies whether the MAC was generated within an "allowable expiry time," which is computed using the "Fudge Factor".

The "Fudge factor" field specifies the duration of time after the MAC generation time, the message can be considered valid. It is computed by applying a "fudge factor" on the MAC generation time (adding or subtracting a small number of seconds) to allow for clock skew (mismatch) between the MAC generator and verifier hosts.

To have a secure transaction based on TSIG, a sender computes the hash of the entire DNS message and secret key and encodes the result in a TSIG RR at the end of the message. At the recipient end, the TSIG record is stripped from the DNS message and processed. The process whereby the recipient uses the TSIG record to verify the integrity of the received DNS message is called *verification*. The verification process uses the hash algorithm name to identify the hash function and the key name to identify the key to be used to validate the TSIG record. The number in the fudge factor is used to add to or subtract from the signing time to allow for the possible mismatch of clocks of the signer and verifier. Thus, the fudge factor provides the tolerance limit for the MAC validity period computed, based on time of generation.

The purpose of sending the key name in the TSIG record is to enable the verifier (recipient) of the DNS message to use the right key to verify it. It also enables the recipient to verify that the key name is indeed one of the keys shared with the sender. The purpose of the "Time Signed" or timestamp field in the TSIG record is to inform the message recipient about the time of MAC generation. The recipient compares this value with the current clock time at the recipient system to ensure that the MAC was generated within the allowable time specified as part of the TSIG record itself. The purpose of using a timestamp is to prevent replay attacks. For correct verification of generation time against current time, it is essential that the system clocks of the transaction participants be synchronized. Protocols such as the Network Time Protocol (NTP) are available for this purpose.

The verification process consists of the recipient retrieving the appropriate secret key, generating its own hash of the received DNS message, and comparing it with the received hash (in the TSIG record). In this verification process, the receiving name server has performed the following validations:

- The message has been verified as coming from an authenticated source (with whom it shares the secret key).

- The message has not been altered in transit (verified by matching of hash values).

Source authentication counters identify spoofing, and data integrity checking helps to counter corruption and modification of data in transit.

BIND version 8.2 was the first version to introduce TSIG features, and is present in every later version. BIND 9.x's support for TSIG includes features to secure zone transfer and dynamic update transactions[8].

The following operations are needed to set up the environment for enabling DNS transactions to use TSIG:

- The system clocks of the name servers (primary and secondary) participating in DNS transactions must be synchronized (e.g., through NTP). If the primary and secondary servers' clocks differ more than the "fudge factor" described above, the TSIG will fail to authenticate and the zone transfer will fail.

- There should be a secret key generation utility that can generate keys of the required length with sufficient entropy. The key file (the file containing the secret key string) must be securely communicated to the two servers participating in the transaction.

- The key information should be specified in the configuration file through appropriate statements (e.g., key statement and server statement in named.conf configuration file of BIND 9.x).

The key generation process is described in Section 8.2.1. The commands needed to define the keys and instruct the name server to use those keys for all DNS transactions are outlined in Sections 8.2.2 and 8.2.4, respectively. The set of checklists for key file creation and key definition within the name servers is given in Section 8.2.5. Protection of zone transfer transactions and dynamic update transactions using HMAC as specified in TSIG are covered in sections 8.2.7 and 8.2.8 respectively.

---

[8] Some server software such as Microsoft Windows Server 2008 does not implement TSIG, but use lower level transaction security (such as IPSec). To set this up, see http://technet.microsoft.com/en-us/library/ee649243(WS.10).aspx

8.2.1   Key Generation

To enable zone transfer (requests and responses) through authenticated messages, it is necessary to generate a key for every pair of name servers. The key also can be used for securing other transactions, such as dynamic updates, DNS queries, and responses. The binary key string that is generated by most key generation utilities used with DNSSEC is base 64 encoded. The program that generates the key in BIND 9.x is dnssec-keygen. An example of a command that generates a secret key (as opposed to other types of keys, such as public keys, which this program also can generate) by invoking the dnssec-keygen program is as follows:

```
dnssec-keygen -a HMAC-SHA256 -b 112 -n HOST ns1-ns2.example.com.
```

where the various command options (parameters) denote the following:

>-a option: the name of the hashing algorithm that will use the key (HMAC-SHA256 is preferred, but may not be available in older implementations. Use of HMAC-SHA1 is allowed, but migration to HMAC-SHA256 should be done when available)

>-b option: the length of the key (here – a minimum of 112 bits)

>-n option: the type of key (in this case, the HOST)

>last parameter: the name of the key (ns1-ns2.example.com)

The dnssec-keygen program generates the following files, each containing the key string:

```
Kns1-ns2.example.com.+157+34567.key
Kns1-ns2.example.com.+157.34567.private
```

When the program is generating a pair of keys (one public and the other private), the file with the extension *key* will contain the public key string and the file with extension private will contain the private key. Because in this case only the secret key is being generated, the key strings in both files will be the same for the TSIG implementation. The key string from any of these files is then copied to a file called the key file. This file is then referenced using an include statement within the key statement.

8.2.2   Defining the Keys in the Communicating Name Servers

The key generated by using the dnssec-keygen utility has to be defined within the named.conf configuration file of the two communicating servers (generally one primary name server and one secondary name server). This is accomplished by using the key statement of BIND:

```
key "ns1-ns2.example.com." {
     algorithm hmac-sha256;
     include "/var/named/keys/secretkey.conf";
};
```

where the file secretkey.conf will contain the keyword secret and the actual key string (in this example):

```
secret "MhZQKc4TwAPkURM==";
```

8.2.3   Defining the Keys in a NSD Configuration File

In NSD, declaring a TSIG key is very similar to the example above, with some minor syntax changes:

```
key:
        name: ns1-ns2.example.com.
        algorithm: hmac-sha256
        secret: "MhZQKc4TwAPkURM=="
```

8.2.4   Instructing Name Servers to Use Keys in All Transactions

The command to instruct the server to use the key in all transactions (DNS query/response, zone transfer, dynamic update, etc.) is as follows:

```
server 192.249.249.1 {
        keys { ns1-ns2.example.com.; };
```

The same statement can be used as an entry in an `acl` statement as well:

```
acl key_acl {

        ns1-ns2.example.com.; };
```

8.2.5   Checklists for Key File Creation and Key Configuration Process

Checklist item 8: The TSIG key should be a minimum of 112 bits in length if the generator utility has been proven to generate sufficiently random strings [800-57P1]. 128 bits is recommended.

Checklist item 9: A unique TSIG key should be generated for each pair of communicating hosts (i.e. a separate key for each secondary name server to authenticate transactions with the primary name server, etc.).

Checklist item 10: After the key string is copied to the key file in the name server, the two files generated by the dnssec-keygen program should either be made accessible only to the server administrator account (e.g., root in Unix) or, better still, deleted. The paper copy of these files also should be destroyed.

Checklist item 11: The key file should be securely transmitted across the network to name servers that will be communicating with the name server that generated the key.

Checklist item 12: The statement in the configuration file (usually found at /etc/named.conf for BIND running on Unix) that describes a TSIG key (key name (ID), signing algorithm, and key string) should not directly contain the key string. When the key string is found in the configuration file, the risk of key compromise is increased in some environments where there is a need to make the configuration file readable by people other than the zone administrator. Instead, the key string should be defined in a separate key file and referenced through an include directive in the key statement of the configuration file. Every TSIG key should have a separate key file.

Checklist item 13: The key file should be owned by the account under which the name server software is run. The permission bits should be set so that the key file can be read or modified only by the account that runs the name server software.

Checklist item 14: The TSIG key used to sign messages between a pair of servers should be specified in the server statement of both transacting servers to point to each other. This is necessary to ensure that both the request message and the transaction message of a particular transaction are signed and hence secured.

In each of the configuration files of the pair of servers that share a secret key in a zone, the name of the key to be used for all communication between them must be specified (via the server statement in BIND configuration files).

8.2.6   Securing Zone Transfers using TSIG

The pair of servers participating in zone transfer transactions must be instructed to use the key defined using the key statement (Section 8.2.2). This pair generally consists of a primary name server and a secondary name server. The primary name server is configured to accept zone transfer requests only from secondary name servers that send MACs using the named key along with a zone transfer request message. The configuration is accomplished by using the allow-transfer substatement of the zone statement. A sample allow-transfer substatement that specifies that the primary name server should only allow zone transfer requests for the example.com zone from name servers that use the ns1-ns2.example.com key is as follows:

```
zone "example.com" {
     type master;
     file "zonedb.example.com";
     allow-transfer { key {ns1-ns2.example.com.}; };
};
```

The secondary name server is instructed to use the key ns1-ns2.example.com in the zone transfer request to the primary name server (with IP address 192.249.249.1) using the server statement shown in Section 8.2.3.

In NSD, the syntax is similar, but there is a special requirement if no TSIG key is to be used. The zone option "provide-xfer" is used to indicate which IP addresses can request a zone transfer for this zone on this server:

```
zone:
     type: master;
     file "zonedb.example.come";
     provide-xfer: 192.68.0.1 ns1-ns2.example.com.
     provide-xfer:  192.68.0.2 NOKEY
```

8.2.7   Securing Dynamic Updates Using TSIG or SIG(0)

Dynamic update restrictions based on TSIG keys can be specified in BIND 8.2 and later versions[9] by using the allow-update substatement of the zone statement. The arguments to this statement are the

---

[9] As mentioned earlier, Micosoft Server does not use TSIG for dynamic update, but uses IPSec instead. See:
http://technet.microsoft.com/en-us/library/cc753751.aspx

keyword key followed by the name of the TSIG key. (See Section 8.2.2 for details on how to enter in the key statement in a BIND name server configuration file.) Once the key statement has been entered, the following substatement can be added to the zone statement to make use of the secret key for dynamic updates:

```
zone "example.com" {
      type master;
      file "zonedb.example.com";
      allow-update { key dhcp-server.example.com.; };
};
```

Note that although the string dhcp-server.example.com. looks like a FQDN, it actually denotes the name of the TSIG key. The implication of the configuration statement example is that any host that possess the key named dhcp-server.example.com. can submit dynamic update requests (adding, deleting, or modifying RRs) to the zone file (for the zone example.com) that resides in the primary authoritative name server.

To use SIG(0) to authenticate dynamic update messages, the key used must first have its public component stored in the DNS, so a validating client can obtain it[10]. See section 9.5 on publishing keys. The previous steps above need to be performed to control access (if desired). After that is done, the updating name server (i.e., primary authoritative name server) should be able to obtain the key and process the dynamic update request (if the name server supports SIG(0) with dynamic update).

8.2.8 Configuring Dynamic Update Forwarding Restrictions Using TSIG Keys

Dynamic updates are allowed on the copy of the zone file in the primary authoritative name server only because that is the only "writable" copy. This does not automatically imply that the primary authoritative name server is the only one allowed to accept dynamic update requests. In fact, BIND 9.1.0 and later versions allow secondary name servers to accept dynamic update requests and forward them to the primary authoritative name server. In this scenario, if there are no restrictions on the basis of the identity of hosts from whom the secondary name server can forward such dynamic update requests, it is equivalent to circumventing the dynamic update restrictions specified in the primary name server because the request can literally originate from any host to the secondary name server and be forwarded to the primary name server. To counter this problem, a new substatement, allow-update-forwarding, is now available in BIND versions that have the dynamic update forwarding feature. An example of this allow-update-forwarding statement using TSIG keys is given below:

```
zone "example.com" {
      type slave;
      file "backupdb.example.com";
            allow-update-forwarding { key dhcp-
server.example.com.; };
};
```

8.2.9 Configuring Fine-Grained Dynamic Update Restrictions Using TSIG/SIG(0) Keys

The allow-update substatement specifies dynamic update restrictions based on the originators of dynamic update requests (a specific set of hosts identified by IP address or holding a TSIG key) but not the contents of the zone records. To specify dynamic update access (grant or deny) restrictions based on a

---

[10] For more information, see the note from the dynamic update howto: http://ops.ietf.org/dns/dynupd/secure-ddns-howto.html

combination of domain/subdomain names and RR types (A, MX, NS, etc.), BIND 9 and later versions provide the update-policy substatement within the zone statement. The update-policy substatement bases these restrictions on the TSIG key. In other words, the update-policy statement specifies which TSIG keys (or holders of keys) are allowed to perform dynamic updates on which domains/subdomains and RR types within that domain/subdomain.

The general form of the update-policy statement is as follows:

```
update-policy {
      (grant | deny) TSIGkey nametype name [type]
};
```

where the semantics of each of the statement components are as follows:

grant/deny—allow/disallow dynamic update for the combination that follows

TSIGkey—the name of the TSIG key used to authenticate the update request

nametype—can be one the following with the associated semantics:

name—restriction applies to the domain name specified in the following name field

subdomain—restriction applies to subdomains of the domain specified in the following name field

wildcard—restriction applies to the set of domains specified using the wildcard syntax (i.e., *) in the following name field

self—restriction applies to the domain whose name is the same as that in the TSIGkey field (i.e., the domain name whose records are to be updated has the same name as the key used to authenticate the dynamic update request). In this usage, the contents of the name field become redundant but still should be used in the statement (i.e., the name field cannot be left blank)

name—used to specify the name of the domain. The syntax used and the domains it covers are based on the value used in the nametype field (e.g., if subdomain is the value of the nametype field, then all subdomains of the domain name used are being covered under this statement).

type—an optional field that can contain any valid RRType (except the NSEC type) or the wildcard type 'ANY' (ANY stands for all RR types except the NSEC type). If it is missing, it denotes all RRTypes except SOA, NS, RRSIG, and NSEC. It also is possible to put in multiple RRTypes separated by a space (e.g., A NS).

Examples of update-policy statements and their associated semantics are given below.

Suppose there is a domain sales.example.com within example.com and that name server uses a TSIG key that has the same name as its own domain name (i.e., sales.example.com). All dynamic updates from sales.example.com could be restricted to all resource records of that domain within the zone file as follows:

```
zone "example.com" {
      type master;
```

```
        file "zonedb.example.com";
        update-policy { grant sales.example.com. self
sales.example.com.; };
};
```

All dynamic updates from sales.example.com could be restricted to only A and MX RR types of that domain as follows:

```
zone "example.com" {
        type master;
        file "zonedb.example.com";
        update-policy {
           grant sales.example.com. self sales.example.com. A MX; };
};
```

To allow clients with the TSIG key sales.example.com to update all records pertaining to subdomains of NEsales.example.com except the name server records (RR Type NS):

```
zone "example.com" {
        type master;
        file "zonedb.example.com";
        update-policy {
                deny sales.example.com. subdomain NE
sales.example.com. NS;
                grant sales.example.com. subdomain NE
sales.example.com. ANY; };
};
```

For Microsoft Windows, authentication is provided using GSS-TSIG. System administrators of Windows Servers should consult their implementation's documentation on how to integrate secure dynamic update using GSS-TSIG.

8.3 Recommendations Summary

The following items provide a summary of the major recommendations from this section:

- **Checklist item 7:** It is recommended that the administrator create a named list of trusted hosts (or blacklisted hosts) for each of the different types of DNS transactions. In general, the role of the following categories of hosts should be considered for inclusion in the appropriate ACL:

  - DMZ hosts defined in any of the zones in the enterprise

  - All secondary name servers allowed to initiate zone transfers

  - Internal hosts allowed to perform recursive queries.

- **Checklist item 8:** The TSIG key (secret string) should be a minimum of 112 bits in length if the generator utility has been proven to generate sufficiently random strings [800-57P1]. 128 bits recommended.

- **Checklist item 9:** A unique TSIG key should be generated for each set of hosts (i.e. a unique key between a primary name server and every secondary server for authenticating zone transfers).

- **Checklist item 10:** After the key string is copied to the key file in the name server, the two files generated by the dnssec-keygen program should either be made accessible only to the server administrator account (e.g., root in Unix) or, better still, deleted. The paper copy of these files also should be destroyed.

- **Checklist item 11:** The key file should be securely transmitted across the network to name servers that will be communicating with the name server that generated the key.

- **Checklist item 12:** The statement in the configuration file (usually found at /etc/named.conf for BIND running on Unix) that describes a TSIG key (key name (ID), signing algorithm, and key string) should not directly contain the key string. When the key string is found in the configuration file, the risk of key compromise is increased in some environments where there is a need to make the configuration file readable by people other than the zone administrator. Instead, the key string should be defined in a separate key file and referenced through an include directive in the key statement of the configuration file. Every TSIG key should have a separate key file.

- **Checklist item 13:** The key file should be owned by the account under which the name server software is run. The permission bits should be set so that the key file can be read or modified only by the account that runs the name server software.

- **Checklist item 14:** The TSIG key used to sign messages between a pair of servers should be specified in the server statement of both transacting servers to point to each other. This is necessary to ensure that both the request message and the transaction message of a particular transaction are signed and hence secured.

9. Guidelines for Securing DNS Query/Response

Section 6.1.4 describes the DNSSEC approach for protection of DNS query/response transactions through data origin authentication, data integrity verification, and authenticated denial of existence capabilities. This section describes the mechanisms involved in the DNSSEC approach, the operations those mechanisms involve, and a secure way of performing those operations by using checklists. In other words, this section provides guidelines for secure deployment of DNSSEC features through DNSSEC operations that are supported in name server software. These guidelines are covered in Sections 9.3 through 9.7.

To ensure end-to-end protection of DNS query/response transactions, additional protection measures (apart from what DNSSEC specification provides)—such as securing the communication path between local DNSSEC-aware caching/resolving name servers and stub resolvers—are needed. These measures are discussed in Section 9.8.

Zone administrators need to have an understanding of the logic of dynamic updates in the presence of additional RRsets (especially the NSEC RRSet) introduced into the zone file by DNSSEC specification. This logic is explained in Section 9.9.

9.1 Enabling DNSSEC processing in BIND & NSD

Before a DNSSEC signed zone can be deployed, a name server must be configured to enable DNSSEC processing. In BIND, it is done by adding the following line to the options statement in the named configuration file (named.conf)

```
options {
     dnssec-enable yes;
     };
```

After restart, the name server will now perform DNSSEC processing for DNS query/response transactions.

In NSD and Windows Server, this option statement is not necessary. If the zone file contains DNSSEC Resource Records, they are loaded automatically and the server correctly sends DNSSEC enabled responses when signaled by a client query.

> **Checklist item 15:** Name servers that deploy DNSSEC signed zones or query signed zones should be configured to perform DNSSEC processing.

9.2 DNSSEC Mechanisms and Operations

DNSSEC mechanisms involve two main processes: sign and serve, and verify signature. These processes are described below.

9.2.1 Sign and Serve

The first task in this process is to generate digital signatures associated with every RR in the zone file. Instead of generating a signature for every RR, DNSSEC specifies generation of a signature for an RRSet (a set of RRs with the same owner name, class, and RRType). The digital signature and its associated information (ID of the key used, start and expiry dates for the signature, etc.) are encapsulated in a special RR of RRType RRSIG (Resource Record Signature). The actual key string and associated information about the public key that is to be used to verify the signature (in RRSIG) are given in a special RRType DNSKEY. Another RRType, NSEC (Next Secure), is used to list RRTypes (in canonical order [RFC4034]) available for a given domain (owner name), and a signature (RRSIG RR) for that RRType is generated to provide authenticated proof of nonexistence to queries for any nonexistent RRType in that domain. In addition, there is an optional RRType DS (Delegation Signer) in case a zone wants to vouchsafe the authenticity of the public key of its child zone. In other words, the DS RR carries the signature for the RR that contains the (hash) public key of the child zone. The detailed syntax for each of these additional RRTypes introduced by DNSSEC specification is specified in RFC 4034 [RFC4034]. The most important of these is the RRSIG RR because it contains the actual signature string.

The RRSIG RR, like any other RR, contains the owner name, TTL, class, RRType, and RDATA fields. The digital signature and all its associated information are contained within the RDATA field. The layout of the RDATA field in the RRSIG RR, with all subfields, is shown in Figure 9-1. A brief description of each subfield follows.

RRType Covered	Algorithm Code	Labels
Original TTL		
Signature Expiration		
Signature Inception		
Key Tag	Signer's Name	
Encoded Signature		

Figure 9-1. RRSIG RR's RDATA Field Layout

The "RRType Covered" field is the type of RRSet for which this RRSIG holds the signature. The "Algorithm Code" field is a code integer assigned to represent a given cryptographic algorithm used to generate the signature. The "Labels" and "Original TTL" fields are the number of labels (number of labels in the FQDN) and the TTL value of the RRSet this signature covers. The "Signature Expiration" and "Signature Inception" values are absolute time values that span the signature validity period—the time period for which this RRSIG is considered valid for the zone. The "Key Tag" and "Signer's Name" fields are the hash and FQDN of the DNSKEY RR that the client needs to validate the signature, and the final field is the encoded signature itself.

A zone that contains the additional RRs along with the regular RRs is called a *signed zone*. A name server that hosts these signed zones and includes the appropriate signatures (i.e., the corresponding RRSIGs) along with requested RRs in its response is called a *DNSSEC-aware name server*.

9.2.2 Verify Signature

The response coming from a signed zone is called a *signed response*. A resolver that has the capability to verify signatures in a signed response is called a *DNSSEC-aware validating resolver*. Before a resolver can verify the signature associated with RRsets (in the response) of a zone using the public key of the zone (sent along with the response), it has to establish trust in that public key. In DNSSEC this requirement is addressed by having the resolver go through a subprocess called *building a trusted chain*.

In this subprocess, the resolver starts from a known list of trusted public keys (called *trust anchors*) and establishes trust in the public key of the zone in a given response by traversing through a chain of public keys, building the chain by using the DNS name space hierarchy. The trust anchors in a resolver ideally consist of the root public keys (if root servers are DNSSEC-aware) or public keys of zones lower in the hierarchy. The trust anchor list in a resolver is not built through a DNS transaction; it uses an out-of-band mechanism.

The DNSSEC processes described above involve several name server operations and a few resolver operations. The name server operations are as follows:

- DNSSEC-OP1: Generation of public key-private key pair

- DNSSEC-OP2: Secure storage of private keys

- DNSSEC-OP3: Public key distribution

- DNSSEC-OP4: Zone signing

- DNSSEC-OP5: Key rollover (changing of keys)

- DNSSEC-OP6: Zone re-signing.

The resolver operations are as follows:

- DNSSEC-OP7: Trust anchors' configuration

- DNSSEC-OP8: Establishing trusted chain and signature verification.

Name server operations DNSSEC-OP1 through DNSSEC-OP4 and resolver operations DNSSEC-OP7 and DNSSEC-OP8 are performed either prior to DNSSEC deployment or for secure operation (serving signed responses and verification of signatures) and hence are covered in this section. The remaining name server operations (key rollovers and zone re-signing – DNSSEC-OP5 and DNSSEC-OP6 respectively) are performed on a periodic basis after a fully operational DNSSEC deployment and hence are covered in Section 11.

9.3 Generation of Public Key-Private Key Pair (DNSSEC-OP1)

DNSSEC specifies generation and verification of digital signatures using asymmetric keys. This requires generation of a public key-private key pair. Although the DNSSEC specification does not call for different keys (just one key pair), experience from pilot implementations suggests that for easier routine security administration operations such as key rollover (changing of keys) and zone re-signing, at least two different types of keys are needed. One set is called Key Signing Key (KSK). This key (specifically, the private part of the key pair, called KSK-private) will be used only for signing the key set (i.e., DNSKEY RRSet) in the zone file. The other key type is called the Zone Signing Key (ZSK) (whose private part is called ZSK-private) and will be used to sign all RRsets in the zone (including DNSKEY RRSet). An administrative distinction is made between the KSK and ZSK keys by setting the Secure Entry Point (SEP) flag bit in the DNSKEY RR that represents the public part of those keys (in this case, it would be called KSK-public).

The logic behind creation of two types of key pairs is to provide separate set of functions for each key type and thus reduce the overall complexity of tasks involved in key rollovers and zone re-signing.

Accordingly, the KSK (KSK-private) is used to sign the key set (i.e., DNSKEY RRSet) and is the key type (public component – KSK-public) that is sent to the parent to be used for authenticated delegation. This is done by generating a DS RR, using the hash of the child's KSK-public key and generating a corresponding signature (RRSIG RR) using the parent's own ZSK. The KSK (KSK-public) may also be used as a trust anchor (sometimes called the SEP keys) in validating resolvers to establish trust chains for verification of signatures.

The ZSK (ZSK-private) is to be used for signing the entire zone file (all RRsets). The public portion of this key (ZSK-public) will not be sent to the parent and will always remain in the zone.

The decision parameters involved in KSK and ZSK key pair generation are as follows:

- Choice of digital signature algorithm

- Choice of key sizes

- Choice of crypto period (duration for which the key will be used).

The choice of digital signature algorithm will be based on recommended algorithms in well-known standards. NIST's Digital Signature Standard (DSS) [FIPS186] provides three algorithm choices:

- Digital Signature Algorithm (DSA)

- RSA

- Elliptic Curve DSA (ECDSA).

Of these three algorithms, RSA and DSA are more widely available and hence are considered candidates of choice for DNSSEC. In terms of performance, both RSA and DSA have comparable signature generation speeds, but DSA is much slower for signature verification. Hence, RSA is the recommended algorithm as far as this guideline is concerned. RSA with SHA-1 is currently the only cryptographic algorithm mandated to be implemented with DNSSEC although other algorithm suites (i.e. RSA/SHA-256, ECDSA) are also specified. It can be expected that name servers and clients will be able to use the RSA algorithm at the minimum. It is suggested that at least one ZSK for a zone use the RSA algorithm.

NIST's Secure Hash Standard (SHS) (FIPS 180-3) specifies SHA-1, SHA-224, SHA-256, SHA-384, and SHA-512 as approved hash algorithms to be used as part of the algorithm suite for generating digital signatures using the digital signature algorithms in the NIST's DSS[FIPS186]. It is expected that there will be support for Elliptic Curve Cryptography in the DNSSEC. The migration path for USG DNSSEC operation will be to ECDSA (or similar) from RSA/SHA-1 and RSA/SHA-256 before September 30th, 2015.

The choice of key size is a tradeoff between the risk of key compromise and performance. The performance variables are signature generation and verification times. The size of the DNS response packet also is a factor because DNSKEY RRs may be sent in the additional section of the DNS response. Because the KSK is used only for signing the key set (DNSKEY RRSet), performance is not much of an issue. Compromise of a KSK could have a great impact, however, because the KSK is the entry point key for a zone. Rollover of a KSK in the event of a compromise involves potential update of trust anchors in many validating resolvers.

As far as the choice of key size for the ZSK is concerned, performance certainly will be a factor because the ZSK is used for signing all RRsets in the zone. In terms of impact, however, it is restricted to just a single zone because the ZSK's usage is limited to signing RRsets only for that zone. This is the justification for allowing 1024 bit RSA keys for use with DNSSEC beyond the USG stop date of 2010. Some network components have been shown to have problems handling large DNS responses. The use of 1024 bit RSA keys is still considered acceptable to compensate for this as long as other rigorous key management practices are in place.

The choice of crypto period (rollover period) is dictated by the amount of work required to compromise the given key. The large size of the KSK implies that the crypto period for that key can be long (usually a year or two). This aids in DNS operations as well, as KSK rollover is more disruptive and requires the zone administrator to interact with their parent zone to update the KSK's DS RR in the parent delegation information. In the case of ZSK, the risk of key guessing is higher of its smaller size. This implies that ZSKs must be rolled over more frequently than KSKs (usually between 1-3 months). Since the ZSK is local to the zone, rolling the ZSK is not as disruptive as rolling the KSK.

In the case of ZSK, the risk of compromise may be greater due to more frequent exposure. If the zone allows dynamic update, the ZSK is often stored on the same server as the zone. This factor, combined with the relatively smaller size of the key, implies that ZSKs must be rolled over more frequently than KSKs (usually between 1-3 months).

In terms of the number of keys of each type (KSK and ZSK) to be generated, a good practice is to generate an extra ZSK in addition to the one that will be used for signing. Hence, the zone administrator should use the key generation program to generate one KSK and two ZSKs during initial deployment of DNSSEC. One ZSK is treated as the active key, and its private part (ZSK-private) will be used for signature generation. The other ZSK (ZSK-public) will be made part of DNSKEY RRSet, but its associated private part (ZSK-private) will not be used for signing RRsets. This additional ZSK will provide a readily available ZSK for immediate rollover in emergency situations such as key compromise and a form of advance notification to validating resolvers that this key is to be the one into which the zone is going to roll over after the current crypto period expires. The mere presence of the key in the DNSKEY RRSet enables validating resolvers to cache and establish trust in the new key so that they can immediately use the key for signature verification as soon as rollover occurs.

The recommended digital signature algorithm suite, key sizes, and crypto periods for the KSK and ZSK keys are given in Table 9-1 [800-57P1]. As with all data authentication keys, this table assumes approved components[11] (hardware or software) and management operations are in place within the organization.

---

[11] For example, for US Federal Government use, all cryptographic components must be FIPS 140 approved. See http://csrc.nist.gov/groups/STM/cmvp/documents/140-1/140val-all.htm for a current list of approved components.

Table 9-1. Digital Signature Algorithms, Min. Key Sizes, and Crypto Periods

Key Type	Digital Signature Algorithm Suite	Key Size	Crypto Period (Rollover Period)
Key-Signing Key (KSK)	RSA-SHA1 (RSA-SHA-256) until 2015	2048 bits	12-24 months (1-2 years)
	ECDSA with Curve P-256 or with Curve P-384	f = 224-255 bits	12-24 months (1-2 years)
Zone-Signing Key (ZSK)	RSA-SHA1 (RSA-SHA-256) until 2015	1024 bits	1-3 months (30-90 days)
	ECDSA with Curve P-256 or with Curve P-384	f = 224-255 bits	12-24 months (1-2 years)

In the above table, the digital signature algorithm suite is given as both RSA-SHA1 and RSA-SHA256. This is because as of the time of writing, RSA-SHA1 is the only algorithm that is both Mandatory for implementations and Approved for use in the Federal Government. However, RSA-SHA1 will be phased out and replaced by RSA-SHA256 within the Federal Government. It is expected that not all software will be updated – especially outside the Federal Government. Because of this, DNS administrators may wish to deploy and use both algorithms for a period of time so DNSSEC client software that does not understand RSA-SHA256 can still get some protection from DNSSEC. The DNS root zone uses RSA/SHA-256 for signing, so deployment of RSA/SHA-256 enabled DNS validators has quickened. It is recommended that new DNSSEC deployments (i.e. initial signing) consider using RSA/SHA-256 or ECDSA, rather than going through the complicated process of algorithm rollover.

The use of RSA in DNSSEC is approved until the year 2015. By this time, it is expected that Elliptic Curve Cryptography (ECC) will be specified in the DNSSEC. USG DNS administrators should plan to migrate to the use of ECDSA (or similar) when it becomes available in DNSSEC components. ECC has an advantage of having the same precieved strength as RSA with a smaller key size. This means that the ZSK can be the same size as the KSK and have a longer cryptoperiod than a 1024 bit RSA ZSK.

9.3.1 Key Pair Generation—Illustrative Example

Every DNSSEC-aware name server implementation should provide a utility program for generating asymmetric key pairs (a public-private key pair). The use of one such program, dnssec-keygen (provided by BIND 9.X), is illustrated below:

```
dnssec-keygen –a algorithm - b bits –n type [options] name
```

where *algorithm* (under –a parameter) can be one of the following:

- RSASHA1

- RSASHA1-NSEC3-SHA1

- DSASHA1

- RSASHA256

- RSASHA512

- ECDSAP256SHA256

bits (key size) for the –b parameter has the following ranges:

- [512..4096] for RSASHA1 and RSASHA256, etc.

- [512..1024] for DSA (must be divisible by 64)

- [] for ECDSA

Note: Even though the key generation utility allows for key sizes smaller than the approved size, US Federal DNS administrators shall only generate keys with approved lengths for use to sign their DNS zone [800-57P3].

type for the –n parameter can be one of ZONE | HOST

name is the owner of the key (usually the domain name in the zone apex).

This command generates two files—one containing the public key and the other the associated private key. The generic names for these files are as follows:

```
K<domain_name>+algorithm_id+Key_id.key
K<domain_name>+algorithm_id+Key_id.private
```

The domain_name is the value of the *name* parameter specified on the command line. Example algorithm IDs are:

003 – DSA

005 – RSASHA1

008 – RSASHA256

The key_id is the unique identifier for the key generated by the program.

For example, to generate a ZSK keyset of length 1024 bits that uses the RSASHA256 algorithm suite for signing the zone example.com, the following command would be issued:

```
dnssec-keygen -a RSASHA256 -b 1024 -n ZONE example.com
```

For this command, the following files containing the private and public keys, respectively, are generated:

```
Kexample.com.+008+28345.private
Kexample.com.+008+28345.key
```

In these file names, 005 stands for the algorithm_id and 28345 is the unique key ID.

In the *.key file, the public key information is expressed in the same syntax as that of a zone file RR. The content of the file Kexample.com.+005+28345.key will be:

```
example.com   IN   DNSKEY   256   3   8   BQFG+KGJ7.......... (Base64 encoded
key string)
```

Hence, the contents of the file containing the public key can be readily added to the zone file (zonedb.example.com) contents by using the following command:

```
cat *.key >> /var/named/zonedb.example.com
```

After the DNSKEY RR containing the public key is added to the zone file, the zone's serial number in the SOA RR must be increased prior to actually signing the zone.

9.4   Secure Storage of Private Keys (DNSSEC-OP2)

The private keys in the KSK and ZSK key pairs should be protected from unauthorized access. If possible, the private keys should be stored offline (with respect to the Internet facing DNSSEC-aware name server) in a physically secure, non-network-accessible machine along with the zone file master copy. The signatures generated by using the private keys should be transferred to the primary authoritative name servers through a load process, using a dynamically established network connection (rather than a permanent network link).

This strategy is not feasible in situations in which the DNSSEC-aware name server has to support dynamic updates. To support dynamic update transactions, the DNSSEC-aware name server (which usually is a primary authoritative name server) has to have both the zone file master copy and the private key corresponding to the zone-signing key (ZSK-private) online to immediately update the signatures for the updated RRsets. The private key corresponding to the key-signing key (KSK-private) can still be kept offline. In this scenario, the following measures must be taken to protect ZSK-private:

- The shell from which the key generation program/utility is invoked should not be accessible except by authorized zone administrators/operators.

- The directory in which the private key files are stored (usually a subdirectory under a directory with the same name as the zone, with the file names in that directory having the name structure K<zonename>.+AlgorithmID+<keytag>.private in BIND) should not be accessible/visible except to authorized zone administrators/operators.

- Sufficient fault tolerance should be provided for the contents of the directory containing the private key files, either by having a mirrored disk or through periodic backups on tape, CD/DVDs, or optical media. The backup media also should be protected against disclosure, tampering, or theft.

- Another strategy is to store the private keys in an encrypted file system.

Checklist item 16: The private keys corresponding to both the ZSK and the KSK should not be kept on the DNSSEC-aware primary authoritative name server when the name server does not support dynamic updates. If dynamic update is supported, the private key corresponding to the ZSK alone should be kept on the name server, with appropriate directory/file-level access control list-based or cryptography-based protections.

9.5 Publishing the Public Key (DNSSEC-OP3) and Setting Up Trust Anchors (DNSSEC-OP7)

The operation of verifying the zone data of a particular zone by a DNSSEC-aware resolver begins with that resolver knowing and trusting the public key of that zone (the one whose data is to be verified) or any of its ancestors (parent and zones above the parent in the DNS tree). If the zone to be verified (e.g., example.com) is secure and its parent (.com) is not, the point of trust begins with the zone itself. If the parent (i.e., the .com zone) is secure (i.e., DNSSEC-aware) and the parent of the parent (i.e., a root zone) is not secure, the starting point of trust is the parent (i.e., .com zone). If a root zone is secure, of course that becomes the origin of trust.

Whatever the starting point of the chain of trust (enterprise level, TLD, or root zone), the public key of that starting point should be made known by the associated name server to the DNSSEC-aware resolvers of interest. These public keys known to DNSSEC-aware resolvers are called trust anchors. Because there is no feature within DNS for third-party authentication of public keys (like X.509 Public Key Infrastructure), the public key has to be distributed through an out-of-band (relative to DNS) process. This distribution may be accomplished through channels such as Web sites or e-mails.

The entries in the trust anchor list of a DNSSEC-aware resolver determine whether a signed response from a zone will be categorized as secure or insecure. As mentioned in section 9.2.2, the resolver has to establish trust in the public key of the zone that has sent the response before it performs signature verification. If that trust cannot be established due to not being able to build a trusted chain using any of the entries in trust anchor list, the response will be tagged as insecure. The reason for not being able to build a trusted chain may be one of several potential reasons: There maybe an intermediate zone signed using an algorithm the end validators does not understand or it may be due to the fact that the DNS namespace contains islands of signed zones as opposed to unbroken hierarchical sequence of signed zones. The effect of each trust anchor in labeling responses for a given mapping of islands of signed zones (Figure 9-2) is given in Table 9-2.

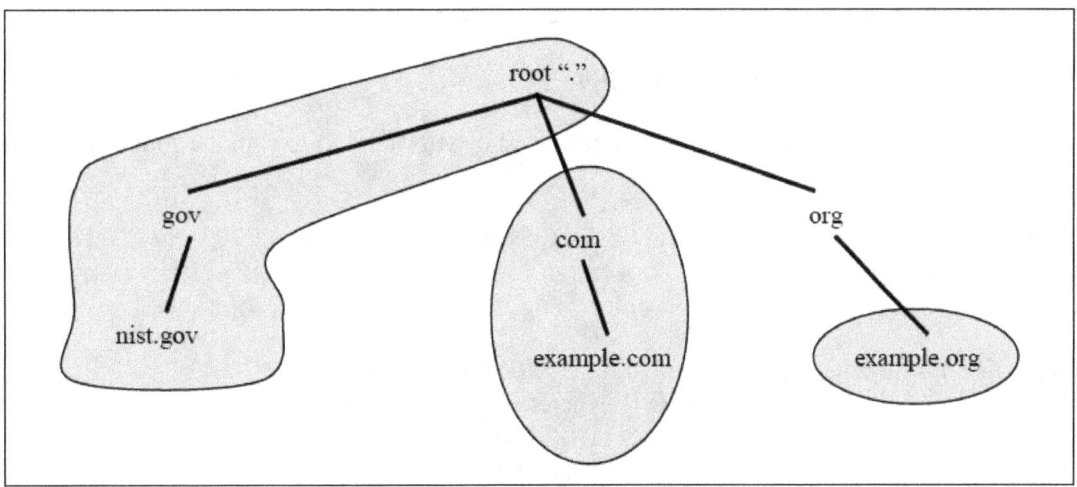

Figure 9-2. A Mapping of Islands of Security

Table 9-2. Impact of Trust Anchor in Labeling Responses

Trust anchor installed on DNSSEC-aware resolver	DNSSEC Response Status for Queries Addressed to		
	www.nist.gov	www.example.com	www.example.org
none	insecure	insecure	insecure
root	secure	insecure	insecure
.com	insecure	secure	insecure
example.org	insecure	insecure	secure

9.6   Zone Signing (DNSSEC-OP4)

The following sequence of actions takes place when a zone file is signed:

- The zone file is sorted in canonical order of domain names.

- An NSEC RR (or NSEC3) is generated for every owner name in the zone.

- The KSK (i.e., KSK-private) is used to sign the DNSKEY RRSet. The KSK is identified by using the SEP flag in the RDATA portion of the DNSKEY RRType.

- The ZSK (i.e., ZSK-private) is used to sign all RRsets in the zone (including DNSKEY RRsets and NSEC RRsets).

Checklist item 17: Signature generation using the KSK should be done offline, using the KSK-private stored offline or using a secure, protected module; then the DNSKEY RRSet, along with its RRSIG RR, can be loaded into the primary authoritative name server.

9.6.1   Zone Signing—Illustrative Example

Use of a zone signer program (dnssec-signzone provided by BIND 9.5.x) to sign the zone is illustrated below:

```
dnssec-signzone -o <name of the zone> -k <name of the file
containing KSK> <location of zone file> <name of file containing
ZSK>
```

To sign the data for the zone example.com with the KSK located in the file Kexample.com.+005+76425.key and the ZSK generated earlier the command would be as follows:

```
dnssec-signzone -o example.com -k Kexample.com.+005+76425.key \
/var/named/zonedb.example.com  Kexample.com.+005+28345.key
```

The process of signing a zone file consists of generating a hash for each RRSet (RRs with the same owner name, class, and RRType), generating a signature for the RRSet (using the private key ZSK-private), and then capturing this signature information in a new RR of type RRSIG RR. The KSK signs just the DNSKEY RRSet, whereas a ZSK signs all RRsets in the zone file. The entity whose private key is used to sign the zone data is called the *signer* or *signing authority*. In most cases, the signer is the domain associated with the zone. In response to a DNS query, a DNSSEC-aware authoritative name server serves the relevant zone data along with the digital signature of the data. The recipient validates the digital signature associated with the received zone data, using the public key of the signer (after establishing trust in that public key). This decryption provides the hash of the original stored zone data. The recipient also generates a hash (using the same hash function algorithm used by the signer) of the received zone data. This computed hash is then compared with the hash retrieved from the digital signature. If the hashes match, then the digital signature in the response is considered validated.

9.7   Establishing Trust Chain and Signature Verification (DNSSEC-OP8)

Until all zones become signed zones, there could be a situation in which a zone is signed but its parent is not signed. The only point of trust (assurance) for the DNSSEC-aware resolver in this scenario is the preconfigured public key of the zone data signer. In the absence of any other source vouching for the authenticity of the public key of the signer, the only source the resolver can trust is the zone signer itself, assuming that the public key is configured to be trusted. On the other hand, if there is a DNS zone X that will vouch for the authenticity of the public key of zone signer child-of-X, the resolver can establish trust in zone signer child-of-X through X. The sequence of zones in the DNS tree through which a resolver establishes trust in the public key of the signer serving signed zone data is called the *chain of trust*.

In the example, the chain of trust starts (originates) with DNS site X and ends with the zone data from site Y. Usually DNS site X will be the zone that is the immediate parent of zone Y (called the *parent zone*) in the global DNS zone tree. The parent zone vouches for the authenticity of the child zone by digitally signing the hash of the public key of the child zone. This hash is stored in a DNS RR called the Delegation Signer (DS) RR. The parent zone also must be a signed zone because a nonsigned zone will not possess the private key (of a private key/public key pair) to sign the hash of the public key of the child zone. Depending on the absence or presence of the chain of trust, a signed zone can be one of the following:

- **Island of security**—a zone that is self-signed. The reason it is self-signed is that the zone's parent is not a signed zone or is unwilling to set up secure delegations and hence cannot vouch for the authenticity of the public key of the child zone. In this instance, there really is no chain of trust.

- **Chained secure zone**—the zone's parent and possibly one or more of the ancestors up the DNS zone tree that are signed. In this hierarchy of signed zones and the associated hierarchy of configured keys, the resolver usually will choose the key at the zone that is highest in this hierarchy of signed zones as the origin of the chain of trust. The public key of the zone at the origin of the chain of trust is called the *trust anchor*. In fact, there may be more than one key that is a trust anchor. In other words, the trust anchors refers to the group of keys that are initially trusted by a resolver, which can then be used to build the chain of trust that leads all the way to the public key of the zone whose signatures the resolver wants to verify. If the chain of trust leads all the way up the DNS tree to the root zone, the zone in question is said to be "globally secure".

Zone data protection through digital signature service for an island-of-security zone consists of the following basic tasks:

- Public key/private key pair generation

- Secure storage (if necessary, offline from the name server) of the private key

- Publication of the public key by incorporation of the DNSKEY RR in the zone file

- Generation of digital signatures for zone data (zone signing).

Additional tasks are involved for a chained secure zone. To be part of the chain of trust, the zone has to inform its parent of its public key (KSK-public) securely (through out-of-DNS channel means). The parent then creates a hash of the public key of its child and stores it in the parent zone in a new RR called a DS RR [RFC4034]. Or the child zone can generate its own DS RRset and upload that to its parent zone for inclusion in the parent zone. It also signs this DS RR by generating a RRSIG RR. Practical conditions dictate, however, that the keys periodically have to be changed because any key can be broken with sufficient computing power. In a chained secure zone, whenever a zone changes its KSK, its parent has to be notified of the new key. The parent then has to add or generate a new DS RR and sign it again. To reduce the administrative burden involved, a common strategy is to use another key pair, the KSK. The KSK is used for signing only the DNSKEY RRSet; all of the other authoritative RRsets in the zone file are signed with the ZSK. The KSK is the key that is published to the parent. The parent will generate (or simply add) the DS RR and a RRSIG RR using the parent's own ZSK. The KSK is used less frequently to only sign the DNSKEY RRset and hence needs to be changed less frequently. There may be situations, however, in which either because of the manageable frequency of key rollovers (key change) or the criticality of DNS information served by the zone, administrators may not use two distinct key pairs for the ZSK and KSK.

In summary, the additional tasks that take place in a chained secure zone are as follows:

- Secure transmission of the KSK or DS by a secure zone to its parent. This transmission is accomplished out of band and may not involve any DNS transactions.

- The parent stores the child's KSK in a special keyset file directory, generates a hash of the key, and stores the hash in a DS RR. The parent also generates the digital signature (RRSIG RR) for this DS RR and includes it with the other delegation information. OR The parent adds the DS to its zone database after the child uploads the DS RRset.

The zone whose signed response needs to be verified (i.e., the target zone) is the leaf node of the chain of trust. The pre-requisite for this operation is to establish trust in the ZSK of the zone. The trust in a zone's ZSK is established through the following operations:

- Authenticated referral from the parent

- Authenticating the KSK of the child zone.

To understand authenticated referral from the parent, it is necessary to look at the normal DNS referral process. In the normal DNS query, a zone that does not have authoritative information pertaining to a query for a domain name in a child zone provides a hint or referral by providing the NS RRSet and associated additional information (RRs that contain the IP addresses for servers provided in the NS RRSet). In normal DNS query processing, following this referral is an acceptable processing step. This process is insufficient, however, from the point of view of establishing a chain of trust; the information about the NS RRSet and associated glue RRs (or any other additional information) cannot be considered authentic because they are not signed with the public key of the authoritative source (i.e., the child zone). To enable authentication of this referral information, the parent provides the cryptographic hint through the DS RR.

Consider an example of how a DNSSEC-aware resolver validates a DNS response from the signed zone example.com. The resolver, following its chain of trust starting from its trust anchor, has authenticated the public key for .com zone and trusts it. Hence, it trusts the NS RRs and DS RR served by the .com zone. From the NS RR and its associated glue RRs, the resolver has determined that the authoritative name server for example.com is ns.example.com and also knows its IP address. Using this information, it goes to ns.example.com and retrieves the KSK of example.com from its DNSKEY RRSet. It hashes this key and compares it with the hash in the DS RR of its parent zone .com. A match of these two hashes authenticates the referral from the .com zone to the example.com zone, as well as the KSK of the example.com zone. Because the ZSK is signed by the KSK of example.com, the resolver can obtain the ZSK of example.com in an authenticated way. Because the ZSK has signed all the RRsets in example.com, any signed response from that zone can be validated by using the now-trusted ZSK.

As the description of the process indicates, the delegation information from a parent in the case of a chained secured zone consists of the following:

- **The NS RRs for the child zone.** Because the authoritative source for these NS records is the child zone, the NS RRs provided are called the hints or referral.

- **The glue RRs.** They provide the location of the servers specified in the NS RRs (IP addresses of referred name servers).

- **The DS RR.** It provides the hash of the KSK or ZSK of the child zone.

- **The RRSIG RR.** It covers the DS RR (signature of the DS RR).

9.7.1 Recording and Communicating Results of Signature Verification

Some name servers can be authoritative for some zones and not authoritative for others. For zones for which they are not authoritative, they perform the function of caching RRsets from previous queries pertaining to those zones. A DNSSEC-aware caching/resolving name server has additional tasks in its caching function. These additional tasks pertain to classifying the security state of the response RRsets and caching them accordingly [RFC4035]. The three possible states of RRsets are as follows:

- **Secure.** A DNSSEC-aware caching name server (or resolver) will categorize an RRSet as secure if verification of its associated RRSIG was successful. This means a valid chain of trust can be formed from the RRSet to a trust anchor. The RRSet would then be placed in the cache and be purged when the data is deemed no longer up to date (according to the TTL) or no longer verifiable (according to the RRSIG validity period).

- **Insecure.** A DNSSEC-aware cache categorizes an RRSet as insecure if it is provably insecure. That means that no DNSSEC security RRs were in the response, and the recipient knew not to expect them. This happens when a delegation to an unsigned zone (a delegation that does not have a DS RR found in the referral response) is received. Insecure RRsets are handled the same way as secure RRsets, but local policy on the end system may dictate not to trust insecure delegations or data in a response.

- **Bogus.** The DNSSEC-aware cache categorizes a response as bogus when verification of the signatures (RRSIG RRs) fails or contains incorrect fields (e.g., RRSIG has expired). The cache policy determines how to deal with these RRsets. They can either be dropped or put into a special BAD cache containing only RRsets that are found to be bogus.

When a DNSSEC-aware caching name server receives a query from a non-security-aware resolver, it serves either secure or insecure data from its cache. Because this query is non-security aware, DNSSEC RR types are not included in the response; so only normal DNS processing applies.

DNSSEC-aware resolvers receive either secure or insecure data from the DNSSEC-aware caching server with a bit set in the header of the DNS response. This header bit is the Authenticated Data (AD) bit, which signals that the RRsets in the response passed all security checks performed by the caching name server. Depending on that server's security posture, the client may choose to accept this response or perform its own set of security checks.

In some cases, a client may want data that has been deemed bogus by a caching name server. In this case, the client sends a query with the Checking Disabled (CD) bit set in the DNS message header [RFC4035]. This tells the DNSSEC-aware caching name server to respond with bogus data from the BAD cache, not an error message. The client system must then perform its own security checks on the response RRsets. In these types of responses, the server does not set the AD bit, indicating that the response has not passed all security checks performed by the server.

9.8   Additional Protection Measures for DNS Query/Response

DNSSEC specifications for protection of DNS query/response transactions cover the following communications:

- DNS responses from remote authoritative name servers to local (enterprise) resolving name servers

- DNS responses from remote caching name servers to local resolving name servers.

Because most of the queries in DNS originate from stub resolvers (on behalf of client software requiring Internet resource access) protection for the DNS response message must be extended to the stub resolver to the resolving name server path as well. The protection approach for this path (also called the *DNS last hop or last mile*) is determined by the nature of the stub resolver and how the network is set up.

Stub resolvers can be non-DNSSEC aware, DNSSEC-aware nonvalidating, and DNSSEC-aware validating. Most of the stub resolvers deployed today are non-DNSSEC aware. In other words, not only do they not have the capability to verify signatures associated with returned RRSets, they also cannot make a distinction between an authenticated (signature verified) response and a nonauthenticated response (passed through by their local resolving name servers). To have complete end-to-end protection for DNS query/response, the minimal requirement for these types of stub resolvers is that they should have the capability to perform origin authentication and data integrity for responses coming down the channel connecting them to the resolving name server providing DNS name resolution service for them. This capability can be installed in these types of stub resolvers with the HMAC approach specified in TSIG (as HMACs are implemented for protecting zone transfer and dynamic update transactions). Alternatively, the enterprise can provide this capability through other network security mechanisms, such as IP Security (IPsec). Guidance on how to set up IPsec protected connections between clients and a recursive caching server is beyond the scope of this document. Whatever mechanism is used for channel security of the last hop (resolving name server to stub resolver), this capability should be present in DNSSEC-aware nonvalidating stub resolvers as well, in addition to non-DNSSEC aware stub resolvers. A DNSSEC-aware nonvalidating stub resolver can leverage this trusted path to examine the setting of the AD bit in the message header of a response message it receives. These types of stub resolvers can then use this flag bit as a hint to find out whether the resolving name server was able to successfully validate the signatures for all of the data in the Answer and Authority sections of the response.

In some situations, establishing a trusted path between stub resolvers and the resolving name servers that provide DNS service for them is not feasible. An example is where the resolving name server is not under the administrative domain of the enterprise but is run by an ISP. In this situation, the end-to-end protection for DNS query/response can be ensured only by having a DNSSEC-aware stub resolver or a stub resolver that can use message authentication (TSIG or SIG(0)) to communicate with a trusted caching server. A DNSSEC-aware stub resolver can indicate to the local resolving name server that it wants to perform its own signature validation by setting the checking disabled (CD) bit in its query messages. Or a stub resolver must be manually configured to use a specific resolving name server and use TSIG/SIG(0) to provide message authentication.

9.9 Dynamic Updates in a DNSSEC-aware Zone

Section 4.3 lists the various logical operations on a zone file during the course of a dynamic update. The four logical operations can be regarded as consisting of two basic operations: addition of RRs and deletion of RRs. Updating an RR can be regarded as a combination of the two basic operations of addition and deletion. Addition and deletion of RRs involve no further operation on the rest of the RRs in the zone file in nonsecure zones. In a secure zone, however, there is one NSEC RR (and a corresponding RRSIG RR) to cover every gap in the namespace.

There is one NSEC RR for every unique owner name in the zone. This NSEC RR points to the next owner name in the canonical order (ordering obtained by lexicographically sorting the domain names within a zone). The NSEC RR for the last owner name in canonical order points to the zone apex name (in other words, the zone name). Hence, conceptually NSEC RRs form a circular link list that traverses the unique domains names in a zone.

Consider the organization and contents of NSEC RRs for the zone example.com. Suppose the following is the canonical order of the unique domain names in the zone:

```
example.com.    IN  SOA  ns.example.com. admin.example.com. (
12985 3600 2700 8000 3600 )
        IN  RRSIG ( SOA )
        IN  NS  ns.example.com.
        IN  RRSIG ( NS )
        IN  MX  mail.example.com.
        IN  RRSIG ( MX )

ns.example.com.  IN  A  192.253.101.7

mail.example.com.  IN  A  192.253.101.8

marketing.example.com.  IN  A  192.253.101.9
        IN  RRSIG ( A )
        IN  MX  mail.example.com.
        IN  RRSIG ( MX )

sales.example.com.   IN  NS  ns.example.com.
        IN  RRSIG ( NS )

www.example.com.  IN  A  192.253.101.10
        IN  RRSIG ( A )
```

The pseudo format (containing only the important fields) of NSEC RRs covering the gaps in the namespace relating to domain names and RR types found at each name in our zone is as follows:

```
example.com.  IN  NSEC  marketing.example.com. (NS SOA MX RRSIG
NSEC)

marketing.example.com.      IN  NSEC  sales.example.com. (A MX
RRSIG NSEC)

sales.example.com.  IN  NSEC  www.example.com. (NS RRSIG NSEC)

www.example.com.  IN  NSEC  example.com. (A RRSIG NSEC)
```

Note that there are as many NSEC RRs as the number of unique domain names in the zone. The NSEC RR whose owner name is example.com points to the next domain in the canonical order (i.e., marketing.example.com). The same is true for NSEC RRs pertaining to the domains marketing.example.com and sales.example.com. The NSEC RR pertaining to the last domain name (i.e., www.example.com) points to the first domain name in the zone (i.e., example.com).

When a query for "package.example.com IN A" arrives (which does not exist in the zone), the authoritative server replies with the NSEC RRSet that proves that the name does not exist in the zone. In this case, the response from the server will consist of the normal DNS reply indicating that the name does not exist and:

- marketing.example.com. NSEC RR indicating there are no authoritative names between "marketing.example.com." and "sales.example.com"

- www.example.com. NSEC RR (the last domain in the zone) proving that there are no wildcard names in the zone that could have been expanded to match the query [RFC4035].

- Accompanying RRSIG RRs for each of the foregoing NSEC records for authentication.

The modifications to the NSEC RRs that are required for the following two operations are as follows:

- Adding a new RR type to an existing domain

- Deleting an RR type from an existing domain

- Adding a new domain name to the zone

- Deleting a domain name from the zone.

Adding a new RR type to an existing domain:

Suppose a new mail host is added to www.example.com. This change will require addition of an MX RR to this domain name. Hence, the NSEC RR for www.example.com must be modified to read as follows:

```
www.example.com.  IN  NSEC  example.com. (A MX RRSIG NSEC)
```

Deleting an RR type from an existing domain:

Suppose the enterprise example.com decides that it no longer needs a separate set of e-mails for people in the marketing department. This change will require removal of the mailserver (RRType = "MX") from the domain marketing.example.com. The modified NSEC RR will now read as follows:

```
marketing.example.com.  IN  NSEC  sales.example.com. (A RRSIG
NSEC)
```

Adding a new domain name to the zone:

To enable customers to order goods online, the enterprise decides to add a separate domain websales.example.com. A separate name server and a set of new hosts also will be added. This new domain requires the following changes to the zone file:

- A new NSEC RR (or NSEC3 RR) for the newly added domain. In this case, a NSEC RR for the domain name websales.example.com should be added. Its location in the canonical order must be determined. This NSEC RR should be set to point to the next domain name in the new canonical order. The presence of the name server and hosts should be reflected in this new RR (through the NS and A codes in the RR types list field) as shown below, and the associated RRSIG RR must be generated. Also note that the new delegation is unsecure, since there is no DS RR indicated in the RR list. Any client looking in the domain should not expect DNSSEC information in replies.

```
websales.example.com.  IN  NSEC  www.example.com. (A NS RRSIG
NSEC)
```

- The NSEC RR (or NSEC3 RR) pertaining to the domain name just preceding the added domain (in the canonical order) has to be modified to point to the newly added domain. The NSEC RR should now point to the domain websales.example.com. A new RRSIG RR must be generated (signed by the ZSK) covering the newly generated NSEC RR.

```
sales.example.com.  IN  NSEC  websales.example.com. (A RRSIG
NSEC)
```

Deleting a domain name from the zone:

Suppose the enterprise has decided that all sales will be made through the Web only. Because hosts have been created to handle this function in the new domain websales.example.com, the hosts in the domain sales.example.com are no longer needed because usage of the applications that reside on them has been discontinued. This implies that all RRs belonging to a domain should be removed from the zone. This change involves the following operations:

- The NSEC RR (or NSEC3 RR) corresponding to the deleted domain should be removed, so the NSEC RR pertaining to domain sales.example.com has to be deleted.

- The NSEC RR (or NSEC3 RR) pertaining to the domain name just preceding the deleted domain has to be modified to point to the domain name immediately following the deleted domain (i.e., pointed to by the deleted NSEC RR). In this case, the NSEC RR for marketing.example.com should point to the domain websales.example.com as follows:

```
marketing.example.com.  IN  NSEC websales.example.com. (A MX
RRSIG NSEC)
```

9.10 Recommendations Summary

The following items provide a summary of the major recommendations from this section:

- **Checklist item 15:** Name servers that deploy DNSSEC signed zones or query signed zones should be configured to perform DNSSEC processing.

- **Checklist item 16:** The private keys corresponding to both the ZSK and the KSK should not be kept on the DNSSEC-aware primary authoritative name server when the name server does not support dynamic updates. If dynamic update is supported, the private key corresponding to the ZSK alone should be kept on the name server, with appropriate directory/file-level access control list-based or cryptography-based protections.

- **Checklist item 17:** Signature generation using the KSK should be done offline, using the KSK-private stored offline; then the DNSKEY RRSet, along with its RRSIG RR, can be loaded into the primary authoritative name server.

10. Guidelines for Minimizing Information Exposure Through DNS Data Content Control

The DNS security extensions provide only origin authentication and data integrity protection. This protection does not provide confidentiality because of the design decision to keep DNS data public. Features such as split DNS provide some means of keeping internal network information from appearing on the global Internet, but they are not official parts of the DNS protocol specification.

It is possible that an attacker may learn valuable information about an organization's network through DNS and use that information to launch an attack. For some elements, such as IP addresses of public servers, this possibility is unavoidable. There are steps a DNS administrator can take in generating a zone file to keep network exposure to a minimum, however; these steps are given below. This process should be done prior to signing a zone. Network information that should be kept absolutely private should not be published in DNS at all.

10.1 Choosing Parameter Values in SOA RR

The first action a DNS administrator should undertake is to ensure that the values in the SOA resource record data are correct. The values in this RR regulate the communication between primary and secondary servers for the zone, such as when secondary servers should periodically perform zone transfers from the primary server. This data also contains the *minimum TTL value* field, which gives client resolvers an indication of how long to cache non-DNSSEC negative responses. Some suggested values for these fields are as follows [RFC1912]:

- **Serial Number.** The *serial number* in the SOA RDATA is used to indicate to secondary name servers that a change to the zone has occurred and a zone transfer should be performed. It should always be increased whenever a change is made to the zone data.

- **Refresh Value.** The *refresh value* tells the secondary servers how many seconds to wait between zone transfers. For zones that are updated frequently, this number should be small (20 minutes to 2 hours). For zones that are updated infrequently, a larger number can be substituted (2 to 12 hours). For a signed zone, this value should not be more than the length of the RRSIG validity period, to ensure that secondary zones do not contain zones with expired RRSIGs. This value also may depend on bandwidth constraints on the primary server side. Note that if a primary server issues a NOTIFY message when it is updated, a secondary server will immediately perform a zone transfer and not wait until the refresh value has timed out.

- **Retry Value.** The *retry value* is the period a secondary server should wait before attempting to perform a zone transfer if the previous attempt failed. This value should be a fraction of the refresh value. If using a Refresh value in the range given above, the possible range of values for this field would be 5 minutes to 1 hour.

- **Expire Value.** The *expire value* is the length of time a secondary server should consider the zone information valid if it can no longer reach the primary server to refresh. This field allows secondary servers to continue to operate until network disruptions are resolved. This value depends on the frequency of changes to the zone and the reliability of the connection between name servers. It should be a multiple of the refresh value and possibly set to as long as 2 to 4 weeks.

- **Minimum TTL.** The *minimum TTL value* is the default value used as the TTL of zone RRsets as well as in negative caching. The minTTL value depends on how often the information changes in the zone. If a zone is static, the value could be large; if the zone is dynamic, the value should be small. A minimum value of 5 minutes is advised, with a recommended range of 30 minutes to 5 days.

Checklist item 18: The refresh value in the zone SOA RR should be chosen with the frequency of updates in mind. If the zone is signed, the refresh value should be less than the RRSIG validity period.

Checklist item 19: The retry value in a zone SOA RR should be 1/10th of the refresh value.

Checklist item 20: The expire value in the zone SOA RR should be 2 to 4 weeks.

Checklist item 21: The minimum TTL value should be between 30 minutes and 5 days.

10.2 Information Leakage from Informational RRTypes

There are several types of RRs in the DNS that are meant to convey information to humans and applications about the network, hosts, or services. These RRs include the Responsible Person (RP) record, the Host Information (HINFO) record, the Location (LOC) record, and the catch-all text string resource record (TXT) [RFC1035]. Although these record types are meant to provide information to users in good faith, they also allow attackers to gain knowledge about network hosts before attempting to exploit them. For example, an attacker may query for HINFO records, looking for hosts that list an OS or platform known to have exploits. Therefore, great care should be taken before including these record types in a zone. In fact, they are best left out altogether.

More careful consideration should be taken with the TXT resource record type. A DNS administrator will have to decide if the data contained in a TXT RR constitutes an information leak or is a necessary piece of information. For example, several authenticated email technologies use TXT RR's to store email sender policy information such as valid email senders for a domain. These judgments will have to be made on a case-by-case basis.

Checklist item 22: A DNS administrator should take care when including HINFO, RP, LOC, or other RR types that could divulge information that would be useful to an attacker, or the external view of a zone if using split DNS. These RR types should be avoided if possible and only used if necessary to support operational policy.

Checklist item 23: A DNS administrator should review the data contained in any TXT RR for possible information leakage before adding it to the zone file.

10.3 Using RRSIG Validity Periods to Minimize Key Compromise

The best way for a zone administrator to minimize the impact of a key compromise is by limiting the validity period of RRSIGs in the zone and in the parent zone. This strategy limits the time during which an attacker can take advantage of a compromised key to forge responses. An attacker that has compromised a ZSK can use that key only during the KSK's signature validity interval. An attacker that has compromised a KSK can use that key for only as long as the signature interval of the RRSIG covering the DS RR in the delegating parent. Therefore, it makes sense to keep these validity periods short, which will require frequent re-signing.

To minimize the impact of a compromised ZSK, a zone administrator should set a signature validity period of 1 week for RRSIGs covering the DNSKEY RRSet in the zone (the RRSet that contains the ZSK and KSK for the zone). The DNSKEY RRSet can be re-signed without performing a ZSK rollover, but scheduled ZSK rollover should still be performed at regular intervals.

To prevent the impact of a compromised KSK, a delegating parent should set the signature validity period for RRSIGs covering DS RRs in the range of a few days to 1 week. This re-signing does not require frequent rollover of the parent's ZSK, but scheduled ZSK rollover should still be performed at regular intervals.

Checklist item 24: The validity period for the RRSIGs covering a zone's DNSKEY RRSet should be in the range of 2 days to 1 week. This value helps reduce the vulnerability period resulting from a key compromise.

Checklist item 25: A zone with delegated children should have a validity period of a few days to 1 week for RRSIGs covering the DS RR for a delegated child. This value helps reduce the child zone's vulnerability period resulting from a KSK compromise and scheduled key rollovers.

Following checklist items 24 and 25 in a particular organization depends on the content management and DNSSEC tools in place and local policy regarding content management. A DNS administrator may be forced to choose one signature validity period for all of the content in a zone if the available tools do not allow for fine grain control in zone signing.

In these cases it may be necessary for a DNS administrator to choose a single signature valididity period for the entire zone. In these cases, the DNS administrator should choose a value that minimizes the risk should a key compromise or emergency update be necessary.

10.4 Hashed Authenticated Denial of Existance

One of the side effects of the DNSSEC security extensions as they were first specified is the ability for a user to "walk" a zone by sending a series of queries for NSEC RRs (See Section 9). In that section the NSEC RR and its use in authenticated denial of existence is introduced. A client could send a query for an NSEC RR in the zone, then "walk" the zone by sending a follow-up query for the NSEC RR at the name indicated as the next name in the received NSEC RR. This would result in a client being able to enumerate the entire contents of a zone. While this is not an attack by itself (all DNS data is considered public), it would most likely be a prelude to an attack. An attacker would enumerate a zone to discover the IP addresses of servers to attack directly.

There are several operational means to minimize the impact of zone enumeration and a protocol option in DNSSEC [IEEE]. The operational options are to use split DNS or some other means of segregating public servers from internal hosts. One way of doing that is described in Section 7.2.8 and Section 7.2.9.

The protocol means to minimize information leakage through zone enumeration is to deploy the NSEC3 variant of the NSEC RR [RFC5155]. The NSEC3 RR is similar in format to the NSEC RR, but uses hashed domain names for the owner name and next name (e.g. using SHA-1 or another one-way hash algorithm). Like RRSIG and NSEC RRs, NSEC3 RRs are generated during the zone signing process, however both servers and clients may need to perform hash calculations during run time to construct responses (on the server side) and validate responses (on the client side). Because of this, NSEC3 deployment requires more computation than DNSSEC signed zones containing NSEC RRs.

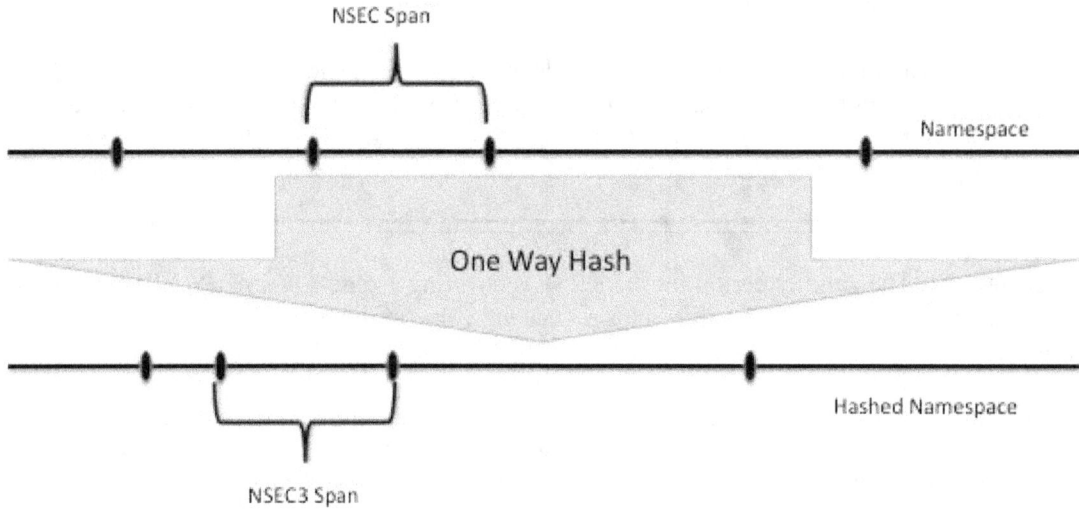

Figure 10-1. NSEC3 Hash Name Space

NSEC3 RRs contain other options than just the (hashed) next name and RRType bitmap. There are also 2 values associated with the NSEC3 RR: the iterations (number of times each name is hashed) and the salt (string appended to each name before hashing). These values are configurable during signing and are used to increase the work necessary by an attacker. Both values should be changed on a regular basis to maintain protection against zone enumeration.

The salt value should be changed every time the entire zone is resigned. The salt value should be a random string with a length small enough to ensure that appending the salt value to the domain name does not result in a FQDN considered too long for the DNS protocol (a single label in the DNS protocol can be 256 octets). A value between 1 - 15 octets would be acceptable for the majority of cases. Note that for zones that are dynamically resigned as needed may not be able to change the salt for NSEC3 RR's as an atomic process. In these cases, the salt rollover procedure is similar to the key algorithm rollover procedure in that the NSEC3 RR chain with the new salt is generated first (ending with the NSEC3PARAM RR) before removing the old (outgoing) NSEC3 chain.

The iterations value depends on factors that are largely beyond the control of DNS administrators. The iterations value is included to increase the amount of work an attacker must complete when conducting a brute force attack offline against a known hashed name from the zone. The goal is to set this value so the amount of work is equal to the time needed to conduct a brute force attack against the server itself (including network lag, etc.). So choosing a value for the iterations requires a good estimate of the computational power available to an attacker and the time required to perform a brute force dictionary attack against an online DNS server. However, setting the value too high would result more work being done by the server and valid clients that may result in slow response time for end users and may be a possible Denial of Service (DoS) vector for servers and clients.

The iterations value should be reviewed annually and changed if conditions warrant. Since the average DNS query response time is in the range of 25 to 200 ms, the iterations value should be chose to reflect the number of hash calculations the average processor could do in that time period. However, in practice the value should be lower to account for client processing power and to speed response for valid clients. Basic empirical work has shown that the ideal iteration value when using SHA-1 should be between 1 and 500, with values 1-200 recommended for average zones. Anything over 500 iterations, it becomes more expensive than validating a signature generated by a 2048 bit RSA key for valid clients. If SHA-256 is used, the value should be between 1 and 300, again with values between 1 and 100 for the majority of zones in practice, as it takes more time to perform a SHA-256 calculation than a SHA-1 calculation.

The OPT-OUT option is designed to aid deployment in very large zones with a very large number of delegations (signed and unsigned). With the OPT-OUT option, the NSEC3 span only links signed delegations, so there are no NSEC3 RRs for names of unsigned delegation. This means there would be a shorter NSEC3 chain, as not every name has an NSEC3 RR associated with it, and the zone signing should take less time. Also, if there are frequent changes, the NSEC3 chain may not have to be changed if unsigned delegations are added or deleted. The drawback is there may be an NSEC3 span an attacker can use to spoof the existence of new, unsigned zones. However, NSEC signed zones can also have spoofed unsigned delegations, but the attacker is limited to re-directing existing names in the zone, while an attacker could spoof completely new unsigned delegations in a zone signed with NSEC3 and using OPT-OUT.

Checklist Item 26: If the zone is signed using NSEC3 RRs, the salt value should be changed every time the zone is completely resigned. The value of the salt should be random, and the length should be short enough to prevent a FQDN to be too long for the DNS protocol (1 to 15 octets should be sufficient).

Checklist Item 27: If the zone is signed using NSEC3 RRs, the iterations value should be based on available computing power available to clients and attackers. The value should be reviewed annually and increased if the evaluation conditions change. Initial values should be between 1 and 200 iterations when using SHA-1 and 1 and 100 if using SHA-256.

10.5 Resource Record TTL Value Recommendations

As mentioned in Section 3.1, each RRset in a zone has its own Time-to-Live value (TTL). This TTL tells clients how long (in seconds) it should be stored in its cache upon receipt. When a client receives a DNS response to a query and performs all relevant checks, it stores the resulting RRsets in its cache. The client cache decrements the TTL value of each RRset in its cache, and when the TTL for any RRset reaches zero, it is purged form the cache. This prevents caches from growing too large, as well as getting old, possibly incorrect, DNS data out of caches and prevents them from being returned to future queries.

The zone administrator assigns the TTL value for each RRset individually or for the entire zone and different RRsets in the same zone can have different values. The zone administrator should set the TTL value long enough to insure that the RRset will be useful for caches, but short enough that any changes to the RRset will be propagated quickly through the DNS (and old information purged). DNSSEC (Section X.X) signature validity periods should also be taken into consideration as well. TTL values should be a fraction of the validity period of the RRSIG that covers the RRset. DNSSEC aware clients will decrement the TTL value of an RRset it its cache to the signature expiration date if that date is before the projected Time-to-Live. That way, the RRset will be purged before it the signature expires (and will be seen as BOGUS to other DNSSEC validators). However, DNSSEC-unaware clients will not know to do this comparison, so there is the risk that invalid DNSSEC RRsets will be stored in DNSSEC unaware caches.

TTL values should be on the order of hours, with a recommended range of 1800 (30 minutes) to 86400 (1 day). If a zone administrator knows that the DNS data is likely to change frequently, the TTL value should be set low, to insure that old, stale data is purged from client caches. If the zone administrator believes the DNS data will not change frequently, then the TTL value can be set higher, to gain optimal benefit of caching in client systems. Note that some specialized load-balancing scenarios rely on much shorter time periods (60 seconds or less), but for the majority of DNS data, 30 minutes to 24 hours is sufficient. If the data is signed using DNSSEC, the value should always be long enough to insure that the data will not be purged from client caches before those clients have a chance to validate it. Experience has shown that very low TTL values (e.g. 30 seconds and under) can cause problems with DNSSEC validating caches and these values should be avoided for DNSSEC signed RRsets.

Checklist item 28: TTL values for DNS data should be set between 30 minutes (1800 seconds) and 24 hours (86,400 seconds).

Checklist item 29: TTL values for RRsets should be set to be a fraction of the DNSSEC signature validity period of the RRSIG that covers the RRset.

10.6 Recommendations Summary

The following items provide a summary of the major recommendations from this section:

- **Checklist item 18:** The refresh value in the zone SOA RR should be chosen with the frequency of updates in mind. If the zone is signed, the refresh value should be less than the RRSIG validity period.

- **Checklist item 19:** The retry value in a zone SOA RR should be 1/10th of the refresh value.

- **Checklist item 20:** The expire value in the zone SOA RR should be 2 to 4 weeks.

- **Checklist item 21:** The minimum TTL value should be between 30 minutes and 5 days.

- **Checklist item 22:** A DNS administrator should take care when including HINFO, RP, LOC, or other RR types that could divulge information that would be useful to an attacker, or the external view of a zone if using split DNS. These RR types should be avoided if possible and only used if necessary to support operational policy.

- **Checklist item 23:** A DNS administrator should review the data contained in any TXT RR for possible information leakage before adding it to the zone file.

- **Checklist item 24:** The validity period for the RRSIGs covering a zone's DNSKEY RRSet should be in the range of 2 days to 1 week. This value helps reduce the vulnerability period resulting from a key compromise.

- **Checklist item 25:** A zone with delegated children should have a validity period of a few days to 1 week for RRSIGs covering the DS RR for a delegated child. This value helps reduce the child zone's vulnerability period resulting from a KSK compromise and scheduled key rollovers.

- **Checklist item 26:** If the zone is signed using NSEC3 RRs, the salt value should be changed every time the zone is completely resigned. The value of the salt should be random, and the length should be short enough to prevent a FQDN to be too long for the DNS protocol (i.e. under 256 octets).

- **Checklist item 27:** If the zone is signed using NSEC3 RRs, the iterations value should be based on available computing power available to clients and attackers. The value should be reviewed annually and increased if the evaluation conditions change.

- **Checklist item 28:** TTL values for DNS data should be set between 30 minutes (1800 seconds) and 24 hours (86400 seconds).

- **Checklist item 29**: TTL values for RRsets should be set to be a fraction of the DNSSEC signature validity period of the RRSIG that covers the RRset.

11. Guidelines for DNS Security Administration Operations

Section 9 deals with operations for deployment and usage of DNSSEC features for protection of DNS query/response transactions. This section deals with periodic security administration operations (and associated checklists) in a DNSSEC-aware enterprise-level zone and how to perform those operations securely.

11.1 Organizational Key Management Practices

The DNS is an intrical part of an enterprise's infrastructure and DNSSEC operations dealing with key management should be in line with policies covering other data authentication keys maintained by the enterprise. These policies would address such where keys are stored, designated personnel tasked with key management operations, etc. Such an overall security policy is out of scope for this document and depends on the enterprise's current infrastructure and any pre-existing security policies already in place.

NIST Special Publication 800-57 Part 1 and Part 3 [800-57P1], [800-57P3] has general guidance on key management and more specific guidance for specific applications. In these two Special Publications, DNSSEC keys fall under "data authentication" keys because they are used to provide authentication and integrity protection to DNS data.

11.2 Scheduled Key Rollovers (Key Lifetimes)

The keys used for zone signing (ZSK) and key signing (KSK) have to be changed because the keys become vulnerable (liable to be cracked) with enough effort. The compromise of a private key means that any site can spoof the zone by signing a bogus RRSet with the private key, thus defeating the purpose of signing the zone file. Key rollover can take place as a scheduled event (scheduled rollover), or it may take place as a result of an emergency (emergency rollover). Emergency rollover occurs for one of the following reasons:

- The private key of the zone has been compromised.

- The private key of the zone has been lost, and the zone is to be updated before the RRSIGs expire.

In scheduled key rollover, the time period (or frequency of change) after which the keys must be changed is determined by several factors.

- The amount of effort needed to rollover the key, and the potential disruption to the zone or current operation.

- Information security policy within the organization.

- The smaller the size of the private key, the easier it is to crack.

Based on these factors, each zone arrives at a desired frequency for key rollovers for ZSKs and KSK. Recall that the KSK (KSK-private) is used for signing only the DNSKEY RRSet, whereas the ZSK (ZSK-private) is used for signing the entire zone file. Apart from the volume of data, the ZSK also is used much more frequently, as in the following situations:

- When a new RR is added (e.g., a new mail server is added and hence a new MX RR is added the zone file)

- When an existing RR's RDATA has changed (e.g., the IP address of a server has changed and hence the corresponding A RR has to be replaced)

- When the signature has expired for an RRSIG RR.

Because of the volume of data handled and the frequency of usage, the size of the ZSK-private key becomes a factor in overall CPU cycles consumed by the digital signature generation operations. Hence, the ZSK used is often relatively small.

Checklist item 30: The (often longer) KSK needs to be rolled over less frequently than the ZSK. The recommended rollover frequency for the KSK is once every 1 to 2 years, whereas the ZSK should be rolled over every 1 to 3 months for operational consistency but may be used longer if necessary for stability or if the key is of the appropriate length. Both keys should have an Approved length according to NIST SP 800-57 Part 1 [800-57P1], [800-57P3].

The impact of a key rollover on the rest of DNS depends on whether the secure zone is locally secure or globally secure (part of a chain of trust).

For a more detailed discussion of the operational steps involved in a key rollover, see the IETF document on DNSSEC operations [RFC4641]. Note that the two processes described for key rollovers (pre-published and dual-signature) can be used in rolling over the ZSK or the KSK. The recommendations below are based on common practice and minimizing the impact of larger responses on clients.

11.2.1 Key Rollover in a Locally Secure Zone

A zone that is only locally secure will have a ZSK, and possibly a KSK that is configured in client resolvers, as a trusted key. Certain challenges arise when either key is rolled over, although having a KSK even for a locally signed zone makes rolling over the ZSK easier. When a zone changes its ZSK(s) and has a KSK that remains unchanged, the only problem that must be addressed is introducing the new key when the old key may be in some distant resolver's or name server's cache.

The solution is to pre-publish the new public key before the rollover. The DNS administrator needs to publish the new key as a DNSKEY RR in the zone file before it is used to generate signatures. The process is as follows:

- Generate a new key pair.

- Add the public key of the new key pair to the zone file (DNSKEY record).

- Sign the zone using the private key of the currently active key pair and the KSK (if present).

- Wait for a period equal to the TTL of the DNSKEY RRSet or the MinTTL of the zone SOA record (whichever is greater).

- Delete the RRSIG RRs generated by the outgoing key, but retain the DNSKEY RR. Resign the zone using the new ZSK (and current KSK, if used).

- Wait the TTL of the zone's DNSKEY RRset.

- Remove the old, outgoing ZSK from the DNSKEY RRset to reduce the size of the DNSKEY RRset in responses.

- Re-sign the DNSKEY RRSet with the new ZSK.

It might be in the DNS administrator's best interest to perform a ZSK rollover continuously. The administrator can perform the first three steps and wait indefinitely before deleting the old DNSKEY from the key set, even continuing to sign the zone with the old DNSKEY when the RRSIGs in the zone expire. This procedure allows the administrator to perform an emergency key rollover more efficiently (see below).

Zones that pre-publish the new public key should observe the following:

Checklist item 31: The secure zone that pre-publishes its public key should do so at least one TTL period before the time of the key rollover.

Checklist item 32: After removing the old public key, the zone should generate a new signature (RRSIG RR), based on the remaining keys (DNSKEY RRs) in the zone file.

In rolling over the KSK, the secure zone may not know which resolvers have stored the public key as a trust anchor. If the network administrator has an out-of-band method of contacting resolver administrators that have stored the public key as a trust anchor (such as e-mail), the network administrator should send out appropriate warnings and set up a trusted means of disseminating the new trust anchor. Otherwise, the DNS administrator can do nothing except pre-publish the new KSK with ample time to give resolver administrators enough time to learn the new KSK.

11.2.2 Key Rollover in a Chained Secure Zone

A globally secure zone uses two sets of keys: the ZSK and the KSK.

11.2.3 ZSK Key Rollover in a Chained Secure Zone

The operations involved in a ZSK rollover in a globally secure zone are no different from those in a ZSK key rollover in a locally secure zone. (See Section 11.2.)

11.2.4 KSK Key Rollover in a Chained Secure Zone (Manually without Revoke Bit)

The KSK (KSK-public) is the key for which the secure parent of a secure zone provides trust. It provides this trust by using a new RR type called the DS RR, which contains the hash of the child's KSK. The delegating parent signs this DS RR with its own ZSK. When a zone changes its KSK, the trust relationship between itself and its parent will be broken. To maintain this trust relationship, the zone making the KSK rollover must:

- Generate a new KSK and add it to the zone key set

- Sign the zone key set with the new KSK as well as the old KSK (the key about to expire)

- Communicate its new KSK to its parent in a way that the parent can authenticate it

- In addition, the parent has to generate a new DS RR (which will replace the old DS RR) that contains the hash of the new KSK and then sign the newly generated DS RR.

To enable the parent to authenticate the new KSK (referred to as KSK2) based on the existing chain of trust, the zone performing the key rollover (i.e., the child zone) generates a DNSKEY RRSet using the DNSKEY RRs of the existing key together with a new DNSKEY RR for the KSK2. It then generates two signatures (two RRSIG RRs)—one using the private key of the existing KSK and the other using the private key of the new KSK. It then sends the newly generated DNSKEY RRSet together with the RRSIG RRs (in plural because there are two signatures, corresponding to KSK and KSK2, respectively) to the parent. The set of RRs sent to the parent is given below in a pseudo format:

```
example.com   DNSKEY   <key-id: 43543> /* the new KSK */
example.com   DNSKEY   <key-id: 78546> /* the existing KSK */
example.com   DNSKEY   <key-id: 98342> /* ZSK */
example.com   RRSIG(DNSKEY) <signer:example.com signing-key:
78546>  /* the entire DNSKEY RRSet signed with existing KSK */
example.com   RRSIG(DNSKEY)  <signer:example.com signing-key:
43543>  /* the entire DNSKEY RRSet signed with new KSK */
```

On receiving this information, the parent performs the following functions:

- It verifies whether the KSK it trusts (i.e., 78456) has indeed signed the newly generated DNSKEY RRSet that includes the new KSK (KSK2). It does this using the newly generated DNSKEY RRSet, the first of the RRSIG RRs (the one that says 78456 as the signing key,) and its own DS RR.

- It also verifies that KSK2 is authentic (the child zone holds the private key of KSK2) by verifying the second RRSIG RR (the one that says 43543 as the signing key) against the DNSKEY RRSet.

- It then generates a new DS RR containing the hash of the new KSK (KSK2).

- The parent must then generate a RRSIG covering the new DS RR of the KSK2.

With these tasks performed by the parent, the KSK rollover process from the child's perspective is essentially complete. The parent zone must then update its zone file with the new DS RR. This is similar to updating any other RRSet, which includes generating any new RRSIGs, as required.

- The parent zone adds the new DS RR (DS2) to the DS RRset in the delegation

- The parent waits for the new DS RR to propagate (min of the TTL of the DS RRset)

- The parent can then remove the expiring DS RR from the DS RRset, triggering a resigning of the zone

The delegated child zone should keep the expiring KSK in its key set until it has confirmed that the parent has performed the appropriate DS update to ensure that the chain of authentication remains intact during the KSK rollover. After the child zone confirms that the delegating parent has updated the delegation information for the child, the child zone administrator should remove the old KSK from the keyset and re-sign the keyset with the ZSK and the new KSK. The administrator should remove the old (now expired)

KSK to keep the DNSKEY RRset response size from growing too large for a UDP packet. This triggers the use of TCP, which is requires more network resources for the server to respond to clients.

11.2.5 KSK Key Rollover in a Chained Secure Zone (using the Revoke Bit)

There is a standardized method developed in the IETF to provide a stable means to updating a KSK that may or may not be installed in clients as a trust anchor [RFC5011]. Since it is impossible to know if a server's KSK is a trust anchor for a client, it is recommended to use this method to insure that clients update their trust anchors.

This KSK rollover method involves a series of timed steps and changing of flag bits in the KSK DNSKEY RR, which may be difficult to perform manually. There have been several tools developed to automate this process, but the steps are presented below to help administrators understand the process:

1. The administrator adds the new KSK but does not use it to generate any signatures (similar to pre-publishing during a ZSK rollover.

2. Wait a pre-calculated time (recommended – 30 days)

3. Set the Revoke bit (flag) in the outgoing KSK and sign the DNSKEY RRset with both old and new KSK's. The old key is now revoked and will not be used by clients who understand the revoked bit and the process described in RFC 5011. Have the delegating parent add a new DS with the hash of the new KSK.

4. Wait a pre-calculated time (recommended – 30 days)

5. Remove the old KSK and resign using the new KSK only. Contact the delegating parent (if the parent is signed) to delete the corresponding DS RR for the old KSK.

11.3 Emergency Key Rollovers

An emergency key rollover occurs when one or more of the keys in the zone (ZSK or KSK) have become compromised or the private component has been lost and a re-signing needs to occur. These types of rollovers are not planned, so there is a greater chance of a break in the chain of authentication if the zone administrator needs to perform this type of rollover.

11.3.1 Emergency ZSK Rollover

A DNS administrator needs to perform multiple steps to perform an emergency ZSK rollover in the event one (or more) ZSK's are compromised. This multi-step recovery process is another reason for a zone administrator to perform a ZSK rollover continuously. If a zone administrator has the new DNSKEY RR already published in the zone as part of the zone key set (the first three steps in the process listed in Section 11.1.1), the next step depends on the key that has been compromised.

If the ZSK currently in use has been compromised, the zone administrators can immediately rollover to the new key. There is no need to wait for the TTL value to expire because the new key has already been published for a period of time. The administrator can simply strip the old RRSIGs from the zone and re-sign with the new ZSK earlier than the planned rollover. DNS operation should continue normally after this re-signing, with no breaks in the authentication chain.

If the new ZSK (the next ZSK to be used) has been compromised, it should be replaced immediately in the zone key set. This replacement also can be done atomically because the new key was not used to produce any RRSIGs; only the zone key set needs to be re-signed. It also is possible simply to remove the compromised key and replace it with a new ZSK in one update.

There still is a danger of attackers using the compromised ZSK to forge responses coming from the zone. This danger exists as long as the current KSK is still active, with the option of initiating a KSK rollover. Once a ZSK has been compromised, a zone administrator should initiate a rollover of the KSK as soon as possible.

11.3.2 Emergency KSK Rollover

When a zone's KSK has been compromised, the only response is to initiate a KSK rollover with the parent. This rollover, however, is not the same process as a scheduled KSK rollover. There must be a way to alert the parent zone administrator that the old KSK has been compromised and not to accept any KSK rollover messages using that key. The replacement key must be transmitted and verified using some other secure channel to ensure the child zone's identity and may involve one or more non-DNSSEC authentication methods. This process could be the same as that used to establish the original KSK of the child zone.

Checklist item 33: A DNS administrator should have the emergency contact information for the immediate parent zone to use when an emergency KSK rollover must be performed.

Checklist item 34: A parent zone must have an emergency contact method made available to its delegated child subzones in case of emergency child subzone KSK rollover. There also should be a secure means of obtaining the subzone's new KSK.

The only way to minimize the exposure period to a minimum, it is a good idea for a zone administrator to keep the signature validity period (time between signature inception and expiration fields in the RRSIG record RDATA) as short as possible. Since a compromised KSK can only be used by an attacker for as long as the non-compromised parent zone vouches for its security. The downside to shorter signature validity periods is that it leads to frequent re-signing (see below). A zone administrator must take the resources required in re-signing into consideration when choosing a signature validity period for delegation DS RRset RRSIG(s).

11.4 Re-Signing a Zone

The zone file has to be re-signed (new RRSIG RRs generated) in the following situations:

- The signatures have expired or are about to expire.

- The zone file content has changed (as a result of dynamic updates for example).

- One of the signing keys has been compromised or is scheduled for replacement.

There are two strategies for re-signing zone data:

- **Complete Re-Signing.** All existing signature records (RRSIG RRs) are deleted, the zone file is sorted again, all of the NSEC RRs are regenerated, and finally the new signature records are generated. Complete re-signing is performed under the following scenarios:

  - The zone file is generated from a back-end relational database.

  - The zone administrator has set up a batch job to run at a specific date/time (called a cron job in Unix), based on the expiration time set in the signer.

- **Incremental Re-Signing.** NSEC RRs are modified because an existing RRSet has been dropped, NSEC RRs are added because a new RRSet set has been added to the zone file, and signatures are generated only for RRsets that have changed. Incremental re-signing is done when changes to zone file contents have been minimal since the last time of signature generation, as is usually the case after a dynamic update.

Checklist item 35: Periodic re-signing should be scheduled before the expiration field of the RRSIG RRs found in the zone. This is to reduce the risk of a signed zone being rendered bogus because of expired signatures.

Checklist item 36: The serial number in the SOA RR must be incremented before re-signing the zone file. If this operation is not done, secondary name servers may not pick up the new signatures because they are refreshed purely on the basis of the SOA serial number mismatch. The consequence is that some security-aware resolvers will be able to verify the signatures (and thus have a secure response) but others cannot.

11.5 DNSSEC Algorithm Migration

Eventually, it may be necessary to migrate to a new DNSSEC signing algorithm. This may be because of a discovered weakness in the currently used algorithm, or due to overriding policy decisions. The migration from the current DNSSEC algorithm to a new algorithm requires a set of steps and delays. This is due to the need to wait for old data to be removed from caches before the next step.

A proposed process can be found in RFC 4641bis[12], which outlines the basic steps. In order to reduce the risk of a validator thinking it is under a downgrade attack. To reduce this risk, the signatures for the new algorithm are added before the DNSKEY RR with the new algorithm public key. Likewise, when removing the retiring algorithm, the public key DNSKEY RR is removed first, followed by the signatures. The steps outlined in the document are repeated below:

1. The zone starts with public keys of algorithm A and RRSIG RRs generated by algorithm A. The zone administrator then generates a new KSK and ZSK for the new algorithm B.

2. The zone administrator signs the zone with the new algorithm B and adds the RRSIGs generated by algorithm B. There is no need to resign the entire zone (unless algorithm A RRSIGS are due to expire). Only the SOA RR needs to be resigned by algorithm A's ZSK as the serial number has changed (indicating a zone update).

---

[12] As of the time of writing, RFC 4641 is currently undergoing a revision. The working version is draft-ietf-dnsops-rfc4641bis-01.txt

3. Zone administrator waits the TTL of the zone, then adds the new DNSKEY RRs of algorithm B and resigns the SOA RR and DNSKEY RRset.

4. The Zone administrator sends the algorithm B key material (either DNSKEY RR or DS RR) to their parent zone if applicable. If the zone administrator believes they will sign with both algorithm A and B for some time, there should be an agreement that the parent zone maintains both DS RRs (for both algorithms) during this period.

At this point the zone can be maintained as is. When the zone administrator wishes to retire algorithm A and solely sign using algorithm B, the following steps should be taken:

5. The zone administrator asks its parent zone to remove the DS RR generated using the KSK of algorithm A.

6. The zone administrator waits for the TTL of the (now removed) DS. Next, the zone administrator removes the KSK and ZSK of algorithm A from the zone, resigns the SOA RR and DNSKEY RRset (with both algorithm A and B), but maintains the RRSIGs generated by algorithm A over the zone data.

7. The zone administrator waits the TTL of the zone, and then strips the RRSIGs of algorithm A from the zone. Only the SOA RR needs to be resigned at this step, as the rest of the zone data has not changed. The zone is now signed solely with algorithm B at this point.

There are two main concerns an administrator needs to keep in mind when transitioning to a new DNSSEC algorithm. First, during this transition period, response messages from the zone will be larger due to the presence of multiple RRSIGs for every RRset in the zone. During steps 4-6, the DNSKEY RRset for the zone would contain four DNSKEY RRs and four RRSIGs (one from each key in the set). It is not recommended that any other key maintenance operation be done in conjunction with this transition, such as pre-publishing a new ZSK. If the administrator chooses to operate using two signing algorithms for an extended period of time, it is recommended to stagger key maintenance operators so only one key (out of the minimum two KSK's and two ZSK's) should be rolled at a given time.

Secondly, when the transition is complete, the zone may become provably unsecure for some validators. End user validators that do not understand the new algorithm will act as if the zone was provably insecure, as valiadators will not be able to validate RRSIGs generated by the unknown algorithm.

11.5.1 Special Considerations in Key Rollovers with Multiple Signing Algorithms in Use

There is an issue that arises when deploying a new digital signature algorithm with a signed zone. The zone administrators now have to maintain two different key life cycles. There can be a problem when the administrator wishes to pre-publish new ZSK's or KSK's in a zone that is signed by two or more algorithms.

One common rollover regime is to pre-publish the next ZSK the same time as the current ZSK. So there would normally be three DNSKEY RRs in a zone's keyset: one active KSK, one active ZSK, and one pre-published ZSK. So if two algorithms are used, there would be twice that number, or six DNSKEY RRs in the zone's keyset. Once a year (or once every 1 to 2 years), the KSK is also rolled over so there will be a period when a pre-published KSK will also appear in the zone for a total of four DNSKEYs in the zone for each cryptographic suite in use (eight total if two different suites are used).

The solution would be to stagger key rollovers to minimize the number of pre-published keys in the zone (see Figure 11-X using RSA/SHA-1 and RSA/SHA-256 as the old and new algorithms). One way of doing this is to extend the active key lifetime of each ZSK to two months, then alternating the rollover of either cryptographic suite's ZSK each month. At least one key is rolled over every month, but which cryptographic suite is rolled alternates each month. Extending the active key life time to two months from one (more traditional) is considered a larger security risk, since the key is in use longer (and therefore subject to a longer window of attack), but the risk is still not severe enough for the majority of zones.

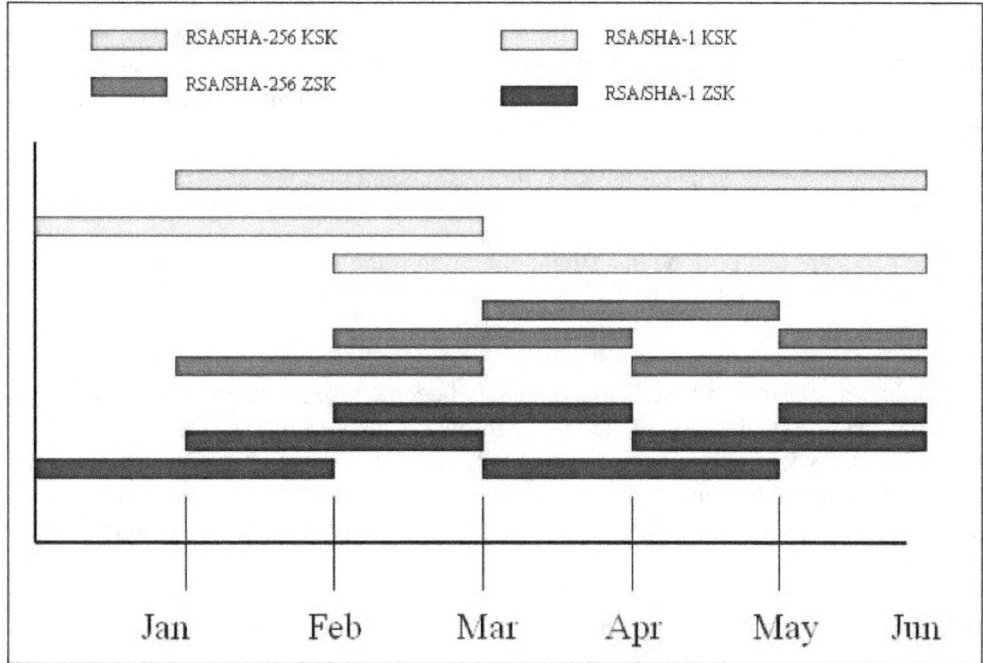

Figure 11-1: Key Lifecycles Using Multiple Algorithms

KSK rollovers can be handled the same way, but since KSK's are rolled over less frequently this is not a big contributor compared to ZSK rollover schedules.

Also note that this process assumes that the zone's KSK is not being used as a trust anchor, and that there is a signed parent for the zone. If that is not case then the zone administrator must follow a variation of the steps in RFC 5011 Section 6.3 [RFC 5011] when finishing the rollover of the KSK. This would involve setting the revoke bit on the algorithm A KSK then waiting the hold-down time before removing it

11.6 Special Consideration When Transitioning from NSEC Signed Zones to NSEC3 Signed Zones

For whatever policy decision, an organization using a DNSSEC signed zone may wish to transition their zone from using NSEC for authenticated denial of existence to using NSEC3 RRs. Due to the nature of caching in the DNS and DNSSEC, this cannot be done in an atomic fashion; i.e., the NSEC chain replaced by an NSEC3 chain in a single operation. This transition is done by moving the DNSKEY RRset from the algorithm code (see Section 9.3) associated with RSA/SHA-1 (code 5) to a special NSEC3 signaling code for RSA/SHA-1 (code 7).

Transitioning is similar to rolling over to a new algorithm since (sometimes) the algorithm code changes as well. The only real difference is at the end when an NSEC3 chain in the signed zone replaces the NSEC chain.

The process is similar to the process described in Section 11.4 above with the following changes:

1. The zone administrator follows the steps 1-5 in Section 11.4 above. The administrator does not have to wait after the NSEC3RSASHA1 KSK appears in the parent zone.

2. The zone administrator waits the TTL of the parent zone and then removes the DNSKEY RRs of the RSA/SHA-1 algorithm. At this point, it is safe to strip the NSEC RRs and replace them with NSEC3 RRs. The DS in the parent zone signals that the zone uses NSEC3 rather than NSEC. As with the algorithm rollover process above, the RSA/SHA-1 RRSIGs over zone data RRsets remain in the zone.

3. The zone administrator waits for the TTL of the zone and then removes the RRSIGs generated using RSA/SHA-1.

Note that the last two steps can be done in a single operation if the zone is small enough. If the zone is too large to sign in a single operation, the resigning procedure defined in RFC 5155 [RFC 5155] in adding NSEC3 RRs using dynamic updates and resigning with NSEC3.

Secondly, when the transition is complete, the zone may become provably unsecure for some validators. This is part of the design consideration of NSEC3 to allow for non-NSEC3 aware validators to revert to a provably unsecure state when encountering NSEC3 RRs rather than generating validation errors. End user validators that do not understand NSEC3 will not recognized the NSEC3 RSA/SHA-1 DNSKEY algorithm code (code 7) and will treat the zone as signed by an unknown algorithm and no validation will be done.

11.7 DNSSEC in a Split-Zone Deployments

11.7.1 Ideal Solution: Internal Delegation

If deploying split horizon DNS for the first time, or if there is a means to re-design an enterprises' DNS, it is recommended that the administrator choose a new internal-only delegation from the primary zone. For example, if the zone were named "example.net", the internal zone would be "internal.example.net".

It may even been that the internal-only zone has an entire separate zone name invisible from the external Internet. For example, an enterprise with the zone "example.net" may choose to have their internal zone be renamed "internal.example." Since the internal network is only accessible by internal hosts, and it is simple to have primary recursive servers act as stealth secondaries for this zone, there would not be an issue of having internal hosts unreachable from inside.

In this case, the network topology would fall along these lines. In the external network:

- Zone file: contains only external (i.e., globally accessible) servers (Web, email, etc.).

- Server configuration: an authoritative-only server with the external zone.

Internal network:

- Zone file: contains internal hosts and any internal interfaces to external servers.

- Server configuration: an authoritative server for internal zone. Or a recursive caching server that also acts as a slave to the internal zone.

11.7.2 Deploying DNSSEC in the Ideal Solution Scenario

In the ideal case (i.e., separate internal zone), DNSSEC deployment can be done independently on both the internal and external zones. The external zone can be signed and have a secure delegation from its parent zone. The internal zone can be signed using a different set of DNSSEC keys (with the name of the internal zone) but does not necessary need a signed delegation from its parent zone (which may not exist). It would be sufficient to simply have the internal public key installed as a trust anchor in all the validators of the organization. This would be similar to managing any other security credentials for the organization.

Since the zones are independent, they can be managed independently. This means that the internal and external zones can have different key rollover schedules if needed. However, as this can complicate operations, it is recommended that the internal and external zones maintain the same rollover schedules to ease operations.

11.7.3 Usual Solution: Same Zone Name with Different Files

In many already existing deployments, it is not possible to create a new subzone or new zone name for internal use. There may be pre-established applications or too many users to make such a transition. In these cases, the DNS administrator for the zone can still deploy DNSSEC on the external and internal zone views, but must take steps in key management to prevent DNSSEC validation failures for internal clients and information leakage to external attackers.

In this scenario, the zone and server configuration should be similar to the following. In the external zone:

- Zone configuration: Similar to the previous scenario. The external zone file (or view) contains only externally accessible hosts such as web, mail and other servers.

- Server configuration: The server either has only the external zone file, or has the zone configured as a separate view that is accessible from the entire Internet.

Internal zone:

- Zone configuration: Again, similar to the previous scenario. The internal zone file (or view) contains only internal hosts and any internally accessible interfaces to external hosts.

- Server configuration: The server either has only the internal zone (or view) or has a separate view with an ACL restricting queries to only internal hosts.

11.7.4 Deploying DNSSEC in the Usual Solution Scenario

DNS administrators need to take some precautions when deploying DNSSEC in this scenario. The administrator can choose to use the same key DNSSEC signing keys for both the internal and external zones, or choose to use separate keys for both zone files. Either choice has its advantages and drawbacks to consider.

If the same key is used for both the internal and external views, the management of they keys are simplified (there is only one set). However, it is easier for an attacker, or mis-configured recursive server to send back name error messages to internal client queries for valid internal hosts. If this happens, a client will receive a valid error message for a host that exists, which will result in a denial of service.

If different keys are used; one set for the internal view and another set for external zones, the zone administrator needs to keep operations separate to prevent using the wrong set for a zone. If this happens, it will appear as if there was an incorrect key rollover, which will result in validation failures.

11.7.5 Considerations for Mobile End Users

The main issues with split horizon DNS deployments come from mobile users moving from the internal LAN to an external network and vice-versa. If the mobile host always relies on an internal recursive server for resolution and validation (via a Virtual Private Network tunnel or some other method), then there is not much that a mobile client needs to be concerned about. The user just needs to be sure that there is a secure connection back to the internal network and an internal recursive server that can accept queries.

It is recommended that mobile hosts have a means to securely access internal hosts via the ability to use an internal recursive DNS server. The client (or administrators) may wish to configure the mobile client to send all DNS queries back through the local internal recursive server if there is any question as to the security of the remote network the mobile system is currently using.

If the client does its own validation (i.e., does not rely on validation being done at the recursive caching server), then it needs to be configured with any internal and external keys as trust anchors. This adds complexity to the DNS management of the end system and may require the end system to change configuration when moving across network boundaries.

It is recommend that a mobile host that does its own validation have the internal DNSSEC key configured as a trust anchor along with any other trust anchors. This would allow the mobile host to perform its own validation even when the local network may not be fully trustworthy. However, when traveling across network boundaries, the mobile client must be able to reconfigure its validator to not use the internal zone key as a trust anchor when not on the internal network.

11.8 Recommendations Summary

The following items provide a summary of the major recommendations from this section:

- **Checklist item 30:** The (often longer) KSK needs to be rolled over less frequently than the ZSK. The recommended rollover frequency for the KSK is once every 1 to 2 years, whereas the ZSK should be rolled over every 1 to 3 months for operational consistency but may be used longer if necessary for stability or if the key is of the appropriate length. Both keys should have an Approved length according to NIST SP 800-57 Part 1 [800-57P1], [800-57P3].

- **Zones that pre-publish the new public key** should observe the following:

  - **Checklist item 31:** The secure zone that pre-publishes its public key should do so at least one TTL period before the time of the key rollover.

  - **Checklist item 32:** After removing the old public key, the zone should generate a new signature (RRSIG RR), based on the remaining keys (DNSKEY RRs) in the zone file.

- **Checklist item 33:** A DNS administrator should have the emergency contact information for the immediate parent zone to use when an emergency KSK rollover must be performed.

- **Checklist item 34:** A parent zone must have an emergency contact method made available to its delegated child subzones in case of emergency KSK rollover. There also should be a secure means of obtaining the new KSK.

- **Checklist item 35:** Periodic re-signing should be scheduled before the expiration field of the RRSIG RRs found in the zone. This is to reduce the risk of a signed zone being rendered bogus because of expired signatures.

- **Checklist item 36:** The serial number in the SOA RR must be incremented before re-signing the zone file. If this operation is not done, secondary name servers may not pick up the new signatures because they are refreshed purely on the basis of the SOA serial number mismatch. The consequence is that some security-aware resolvers will be able to verify the signatures (and thus have a secure response) but others cannot.

12. Guidelines on Securing Recursive Servers (Resolver) & Stub Resolvers

12.1 Setting up the Recursive Service

As discussed earlier in this document, the DNS consists of two basic components: Authoritative name servers that publish DNS data, and DNS resolvers that issue queries for DNS data. Resolvers can be further broken down into two general categories: Stub resolvers (often found on individual hosts) that issue queries, but cannot follow DNS referrals, and recursive resolvers/servers that can follow referrals. The typical set up within an enterprise is to have one or two recursive resolvers/servers providing DNS resolution for all the stub resolvers on the network.

The security threats to an enterprise's recursive resolver/server infrastructure differ slightly from the security threats to its authoritative name servers. Likewise, the measures taken to secure a recursive server differ from an authoritative name server.

12.1.1 Threats to Recursive Servers/Resolvers

Recursive and stub resolvers share a set of threats with authoritative name servers. These threats include the general threats to the host platform (see Section 5.1) and threats to software used to perform DNS transactions (see Section 5.2). However, since recursive (and stub) resolvers do not have zone files (except the root hints file used by recursive resolvers) the administrator only needs to focus on configuration files. The differences would be:

The threats to DNS data contents (Section 5.3) do not directly apply to recursive and stub resolvers as they do not have control over the DNS information published; they only consume it.

12.1.2 Securing the Host Platform for Recursive Servers/Resolvers

Section 7 gives the basic steps in securing the host platform of any DNS component such as running the latest version of software, isolating it on a dedicated server, being aware of known vulnerabilities. Administrators should refer to the relevant Checklist Items from Section 7 with the knowledge that recursive servers do not have zone files (except the root hints file) so there is no need to partition the file as described in Section 7.2.8.

There are additional steps an administrator can take when setting up a recursive server for an enterprise. Since it is unwise to allow queries from the outside Internet to the recursive server, the recursive server can be placed behind a firewall that blocks inbound connections from UDP and TCP port 53 (used by DNS).

Recursive servers should be configured to only accept queries from internal hosts. This is done using the allow-query sub-statement as described in Section 8.1.1.

Checklist item 37: Recursive servers/resolvers should be placed behind an organization's firewall and configured to only accept queries from internal hosts (e.g., Stub Resolver host).

Stub resolvers do not have many configuration options. Often the only configuration necessary for an administrator is to enter the IP addresses of one or more recursive resolvers that the stub would rely on to service queries. It is a good idea for administrators to include at least two IP addresses of recursive servers to increase the availability of the DNS service for end users. This can be done manually, or using a protocol like DHCP.

There is a known class of malware that attempts to change the settings on a system's stub resolver to direct queries to another (usually malicious) recursive server. The server may then direct end users to a malicious site where another attack takes place, or the server may simply direct users to a webpage that serves ads or similar non-intended content. To combat this, administrators should make sure end user systems have the latest anti-virus/malware software and should consider blocking all outbound DNS traffic with the exception of known recursive servers.

Checklist item 38: Administrators should configure two or more recursive resolvers for each stub client on the network.

Checklist item 39: Enterprise firewalls should consider restricting outbound DNS traffic to only the enterprise's designated recursive resolvers.

12.2 Guidelines for Establishing Servers

Enterprise network administrators should set up their recursive server infrastructure with availability and speed of service in mind. However, security is a concern as well as enterprises realize the vulnerability of the DNS to spoofing and importance of monitoring DNS traffic.

12.2.1 Aggregate Caches

Larger enterprises that want to minimize their external network presence may want to consider deploying aggregate caches. An aggregate cache is a recursive resolve that exists on an enterprises' DMZ and only accepts forwarded queries from other recursive resolvers (sometimes called forwarders in this scenario). This reduces the number of servers that need monitoring on the DMZ, but still allows for local recursive servers for separate network segments in an enterprise.

The figure below shows how multiple forwarders send queries to the aggregate cache, which then performs any recursion necessary to find the final answers. The individual forwarders may have their own cache, as well as act as secondaries for internal zones. This is to provide faster service for local stub resolvers.

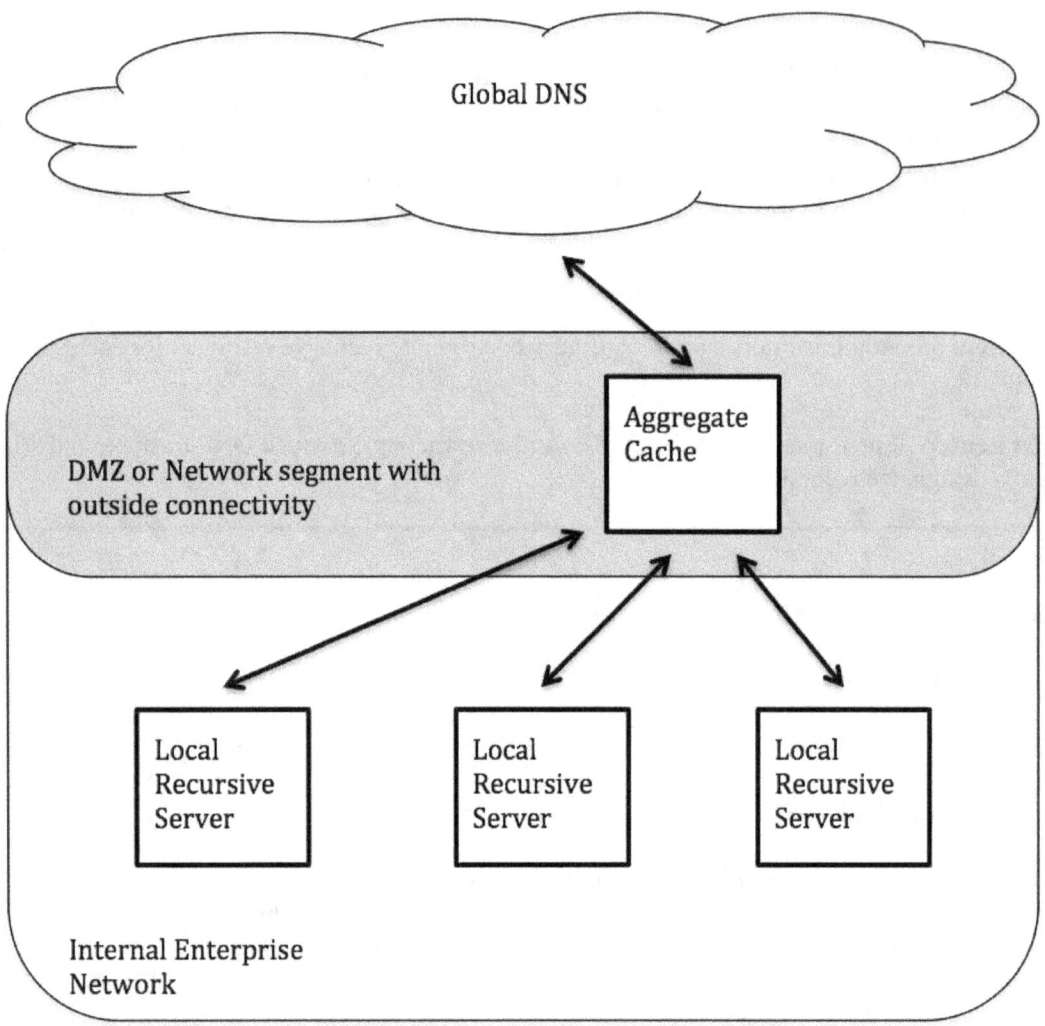

Figure 12-1: Example of Resolver Architecture Using an Aggregate Cache.

When the local forwarder receives a query it cannot answer out of cache (or authoritative data, if present) it forwards the query on to the aggregate cache for recursion. The aggregate cache then performs the recursion and returns the final answer to the forwarder, which in turn sends the final answer back to the stub resolver.

DNSSEC validation on an enterprise with aggregate caches is slightly more complicated than an enterprise that does not use aggregate caches. It is considered wise to perform validation as close to the stub resolver as possible, which in this scenario would mean the forwarder. However, if the aggregate cache does not perform validation, there is the risk of the aggregate cache having DNSSEC invalid data in its cache that would then be sent back to the forwarders. This BOGUS data would only be purged upon the TTL expiration, and not be updated with valid data until that time as passed.

One solution is to perform validation at the forwarder and the aggregate cache, and have the forwarder send all queries to the aggregate cache with the Checking Disable (CD) bit flag set. That signals to the aggregate cache that the forwarder wishes to perform its own validation and should receive the full response back. The aggregate cache is still free to perform its own validation of the data, and the resulting caches on both sides will remain in sync.

> **Checklist item 40**: Whenever Aggregate Caches are deployed, the forwarders must be configured to be Validating Resolvers.

Stub resolvers do not have many configuration options. Often the only configuration necessary for an administrator is to enter the IP addresses of one or more recursive resolvers that the stub would rely on to service queries. It is a good idea for administrators to include at least two IP addresses of recursive servers to increase the availability of the DNS service for end users. This can be done manually, or using a protocol like DHCP.

There is a known class of malware that attempts to change the settings on a system's stub resolver to direct queries to another (usually malicious) recursive server. The server may then direct end users to a malicious site where another attack takes place, or the server may simply direct users to a webpage that serves ads or similar non-intended content. To combat this, administrators should make sure end user systems have the latest anti-virus/malware software and should consider blocking all outbound DNS traffic with the exception of known recursive servers.

> **Checklist item 41**: Administrators should configure two or more recursive resolvers for each stub client on the network.
>
> **Checklist item 42:** Enterprise firewalls should consider restricting outbound DNS traffic to only the enterprise's designated recursive resolvers.

12.3 Setting up the Root Hints File

Every recursive server needs to know the IP address of one or more root servers. If the recursive server cannot find the answer to a stub resolver's query in its cache, it must perform a recursive lookup for the correct information to respond to the client. It does this by first sending the query to one of the root zone servers and following one or more referrals until it reaches one of the authoritative servers of the DNS zone data. See Section 2.2 for a more detailed discussion of the DNS resolution process.

The IP addresses of the root servers are stored in a file often referred to as the root hints file. It is not considered authoritative information, but only hints for the locations of the root zone servers. On startup (and regular intervals), recursive servers send queries to each IP address in the root hints file for the full list of root zone servers. It then maintains this information in its cache until it is time to refresh this data.

It is also necessary for a human administrator to also regularly check the contents of the root hints file for any necessary updates. Missing any changes to the set of root zone servers may result in delays at startup (while waiting for queries to time out) or security vulnerabilities (attackers using spoofed root zone IP addresses). The list of public DNS root zone servers are maintained as part of the IANA functions[13] and administrators can download the current root hints file as well as the DNSSEC public key of the root zone.

---

[13] http://www.iana.org/domains/root/files

Checklist item 43 : Each recursive server must have a root hints file containing the IP address of one or more DNS root servers. The information in the root hints file should be periodically checked for correctness.

Checklist item 44 : The root hints file should be owned by the account under which the name server software is run. The permission bits should be set so that the root hints file can be read or modified only by the account that runs the name server software.

12.4 Securing the Stub Resolver

Stub resolvers do not have many configuration options. Often the only configuration necessary for an administrator is to enter the IP addresses of one or more recursive resolvers that the stub would rely on to service queries. It is a good idea for administrators to include at least two IP addresses of recursive servers to increase the availability of the DNS service for end users. This can be done manually, or using a protocol like DHCP.

There is a known class of malware that attempts to change the settings on a system's stub resolver to direct queries to another (usually malicious) recursive server. The server may then directs end users to a malicious site where another attack takes place, or the server may simply direct users to a webpage that serves ads or similar non-intended content. To combat this, administrators should make sure end user systems have the latest anti-virus/malware software and should consider blocking all outbound DNS traffic with the exception of known recursive servers.

Checklist item 45: Administrators should configure two or more recursive resolvers for each stub resolver on the network.

Checklist item 46: Enterprise firewalls should consider restricting outbound DNS traffic from stub resolvers to only the enterprise's designated recursive resolvers.

12.5 Recommendations Summary

- **Checklist item 37**: Recursive servers/resolvers should be placed behind an organization's firewall and configured to only accept queries from internal hosts (e.g., Stub Resolver host).

Checklist Item 38: Whenever Aggregate Caches are deployed, the forwarders must be configured to be Validating Resolvers.

- **Checklist item 39** : Each recursive server must have a root hints file containing the IP address of one or more DNS root servers. The information in the root hints file should be periodically checked for correctness.

- **Checklist item 40** : The root hints file should be owned by the account under which the name server software is run. The permission bits should be set so that the root hints file can be read or modified only by the account that runs the name server software.

- **Checklist item 41**: Administrators should configure two or more recursive resolvers for each stub resolver on the network.

- **Checklist item 42**: Enterprise firewalls should consider restricting outbound DNS traffic from stub resolvers to only the enterprise's designated recursive resolvers.

- **Checklist item 43** : Each recursive server must have a root hints file containing the IP address of one or more DNS root servers. The information in the root hints file should be periodically checked for correctness.

- **Checklist item 44** : The root hints file should be owned by the account under which the name server software is run. The permission bits should be set so that the root hints file can be read or modified only by the account that runs the name server software.

- **Checklist item 45**: Administrators should configure two or more recursive resolvers for each stub resolver on the network.

- **Checklist item 46**: Enterprise firewalls should consider restricting outbound DNS traffic from stub resolvers to only the enterprise's designated recursive resolvers.

13. Guidelines on Securing Validating Resolvers

As discussed in Section 6.1.4, DNSSEC provides data origin authentication and integrity protection for DNS data. This is done using digital signatures stored with the DNS data and validated by DNS clients. DNS clients that perform this validation are referred to as validators.

DNS validators check responses by building an authentication chain (a.k.a. chain of trust) from the response data to a pre-configured trusted key. This may involve the validator sending follow up queries for public keys or other information (such as DS RR's) to obtain the full DNSSEC chain. Therefore, validators are often recursive servers that can perform their own DNS resolution process and cache results.

13.1 Enabling DNSSEC Validation

Configuring DNSSEC validation requires two tasks: configuring the server to perform validation (not always required) and configure one or more trusted public keys to act as trust anchors. How this is done depends on the DNS implementation used, but BIND will be used as an example.

The policy in determining which DNSSEC public keys to configure as trust anchors is beyond the scope of this guide and would be the part of the organization's security policy. However, given the hierarchy used in DNS, the higher the public key, the wider range of DNS responses can be validated using that key.

13.1.1 On the Stub Client

A stub resolver software, by definition, does not perform full resolution and hence is not DNSSEC-aware. Even if the stub is DNSSEC aware, it may not be able to perform its own validation, but rely on an upstream server to perform validation on its behalf.

13.1.2 Configuring DNSSEC on a Windows 7 or Windows 8 System

Microsoft Windows 7 and Windows 8 operating systems have the option of having the stub resolver to explicitly request DNSSEC validation for a given domain. This is done using the Name Resolution Policy Table (NRPT). A system administrator can set the domain and whether the system should expect DNSSEC validation from responses. These policies can be set on an individual or enterprise wide basis.

The details on how to set this option in Windows is available from the Microsoft TechNet website[14]. Note however that the option is set for an entire domain, not a single zone. If the policy is set for a domain, the client will automatically expect DNSSEC validation for any subzone under the domain as well, and will reject any non-validated responses.

13.1.3 Using a Validating Recursive Server

Most current stub resolvers cannot perform DNSSEC validation and may not understand DNSSEC at all. These stub resolvers will have to rely on an upstream validating recursive resolver to perform DNSSEC validation on its behalf. On an trusted enterprise network, this does not pose a significant risk, as there

---

[14] http://technet.microsoft.com/en-us/library/ee649207(v=WS.10).aspx

are several options to protect the link between a validating recursive resolver and a non-DNSSEC aware stub resolver (see Section 9.8).

If the enterprise network is considered trusted (using one of the last hop mechanisms listed in Section 9.8 or similar), then the stub resolvers can be considered to be using DNSSEC. However, end users will not see DNSSEC validation (especially errors), and network administrators should be aware that DNSSEC validation failures might complicate diagnosis of Internet error messages. DNSSEC validation failures will be seen by the upstream validator and not the stub resolver that initiated the query. The stub resolver will only see a generic server failure message, which applications interpret differently. Network administrators should check validator logs when responding to network errors to rule out DNSSEC validation failures.

Checklist item 47: Non-validating stub resolvers (both DNSSEC-aware and non-DNSSEC-aware) must have a trusted link with a validating recursive resolver.

Checklist item 48: Validators should routinely log any validation failures to aid in diagnosing network errors.

13.1.4 Recommendations for Providing Service to Mobile Hosts

Mobile or nomadic hosts present a particular challenge for network administrators. These systems are often (or always) accessing the network outside of the trusted enterprise; so nomadic hosts must either perform their own validation or have a trusted tunnel back to the enterprise with which to send DNS queries.

Checklist item 49: Mobile or nomadic systems should either perform their own validation or have a trusted channel back to a trusted validator.

If the mobile (or nomadic) hosts can perform their own validation, then the same policy in place for the enterprise validators should be applied for the mobile host. That is, the same trust anchors and validation policy should be set for mobile hosts as for validators on the enterprise networks.

Checklist item 50 : Mobile or nomadic systems that perform its own validation should have the same DNSSEC policy and trust anchors as validators on the enterprise network.

A mobile hosts that can perform its own validation that frequently migrates between the enterpise network and external networks may run into the problem that its local validator may not choose the correct trust anchor for the enterprise zone. In this scenario, the enterprise has an internal and external network with the same zone name (e.g., "example.com") but different zone keys (ZSK and/or KSK). When the mobile host is on the enterprise network, it uses the internal example.com trust anchor to validate DNS responses for internal services. When accessing external facing services, the internal zone trust anchor will not be able to validate the responses about external services and vice-versa (using the external zone key as a trust anchor will result in validation failures for internal DNS zone responses).

In these cases, it may be necessary for a mobile systems to configure its valiadator when migrating from the enterprise to an external network and vice-versa. Or, if the enterprise network is trusted, the mobile host can rely on the enterprise validator when on the enterprise network, and perform its own validation when on external networks. Ideally, network administrators can avoid this problem by using alternative names for internal and external zones, thus having different trust anchors.

If the mobile host cannot perform its own validation, it must have a secure tunnel back to the enterprise network and use an enterprise managed validator. Many enterprises already have a means for mobile hosts to access internal resources (like file servers, etc.); a validating recursive server should be added as one of the services provided to mobile hosts through a secure channel.

13.2 Establishing Initial Trust Anchors

Each DNSSEC validator must have one or more public keys stored locally to use in building authentication chains. These public keys are often referred to as trust anchors. There is nothing particularly special about a trust anchor to anyone other than the client that has configured it as a trusted public key.

Since DNSSEC validation builds upon the DNS hierarchy for trust, the public key for the DNS root zone and TLD's are more valuable because they can (assuming a fully signed DNS domain tree), be used to validate the widest number of nodes (thus a larger portion of the domain name space) and thus provide coverage for a wider range of DNS responses. Validator administrators should identify the set of keys they believe they would need to provide DNSSEC validation for the segments of the name space deemed most important. For most administrators, this would be the DNS root zone key, but may possible include one or more TLD keys (such as the keys for their local ccTLD or similar). Administrators should obtain the public keys (usually the KSK) in a trusted manner (e.g., a SSL secure webpage) and configure the key (or in some cases, a hash of the key) in the implementation specific manner. In BIND, trusted keys are configured using the "trusted-keys" statement in the named configuration file. The format for these keys are:

<zone name>   <flags> <protocol> <algorithm> <base 64 encoded key material>;

Where the <flags>, <protocol> and <algorithm> numbers are taken from the DNSKEY RDATA values. So that in the named configuration file, an example root key would appear as:

```
trusted-keys {
"."    257 3 8 "AwEAAagAIKlVZrpC6Ia7gEzahOR+9W29eux
                hJhVVLOyQbSEW0O8gcCjFFVQUTf6v58fLjw
                Bd0YI0EzrAcQqBGCzh/RStIoO8g0NfnfL2M
                TJRkxoXbfDaUeVPQuYEhg37NZWAJQ9VnMVD
                xP/VHL496M/QZxkjf5/Efucp2gaDX6RS6CX
                poY68LsvPVjR0ZSwzz1apAzvN9dlzEheX7I
                CJBBtuA6G3LQpzW5hOA2hzCTMjJPJ8LbqF6
                dsV6DoBQzgul0sGIcGOYl7OyQdXfZ57relS
                Qageu+ipAdTTJ25AsRTAoub8ONGcLmqrAmR
                LKBP1dfwhYB4N7knNnulqQxA+Uk1ihz0=";
};
```

Since this is the public key, no new security risks are introduced. However, the basic security recommendations for configuration files (see Section 8) should be followed. The risk is an attacker may be able to change the key material to point to a key that the attacker controls. However, this would involve other successful attacks (such as redirecting queries as well) to be fully effective, so it is not considered a high risk.

Checklist item 51 : Validator administrator must configure one or more trust anchors for each validator in the enterprise.

13.3 Maintaining Trust Anchors

Once the trust anchors are initially configured, the administrator must set up a means to monitor that the key is still in use for the give zone, and perform updates as required. Often, zone administrators will perform key rollovers (see Section 11.2) on keys that some clients use as trust anchors. Since there is no means in DNS for servers to contact clients to inform them their trust anchors are out of date, clients must perform their own checks to insure they have an up-to-date list of trust anchors.

The IETF has specified a way for signed zones to signal that a key rollover is underway so clients that have the current zone key can successfully migrate to the new key. This method is specified in RFC 5011 [RFC5011], and some DNSSEC validators that implement this specification can rollover trust anchors in an automated fashion.

Validator administrators should be cogent that not every zone uses the rollover method specified in RFC 5011, and not every implementation can perform automated trust anchor rollovers so regular polling should be done to insure that the validator has the most current key installed as a trust anchor. This could be done via a normal query for the DNSKEY RRset and look for a new key, or by regularly checking the same method the original key was obtained. For example, the DNS root zone key is maintained on website[15] that can be routinely checked (manually or via an automated script) for updates.

How often these checks need to be performed is difficult to determine, but some hints may be found the TTL and RRSIG validity period of the DNSKEY RRset that contains the key used as the trust anchor. Using these values and understanding the timed steps of a key rollover (See section 11.2), a validator administrator can calculate the minimum time between checks.

If a key rollover is detected as in progress, or a new key is anounced, validator administrators should update the validator configuration file with the new key material. This process may be similar to the initial configuration or different, depending on the software used as the validator.

Checklist item 52: The validator administrator regularly checks each trust anchor to ensure that it is still in use, and updates the trust anchor as necessary.

---

[15] https://www.iana.org/dnssec

13.4 Recommendations Summary

- **Checklist item 47**: Non-validating stub resolvers (both DNSSEC-aware and non-DNSSEC-aware) must have a trusted link with a validating recursive resolver.

- **Checklist item 48**: Validators should routinely log any validation failures to aid in diagnosing network errors.

- **Checklist item 49**: Mobile or nomadic systems should either perform their own validation or have a trusted channel back to a trusted validator.

- **Checklist item 50** : Mobile or nomadic systems that perform its own validation should have the same DNSSEC policy and trust anchors as validators on the enterprise network.

- **Checklist item 51** : Validator administrator must configure one or more trust anchors for each validator in the enterprise.

- **Checklist item 52**: The validator administrator regularly checks each trust anchor to ensure that it is still in use, and updates the trust anchor as necessary.

Appendix A—Definitions of Important Terms

Selected terms used in the *Secure Domain Name System (DNS) Deployment Guide* are defined below. The terms come from the DNSSEC specifications [RFC4033], [RFC4034], [RFC4035] and [RFC3833] with some additions for specific terms used in this guide.

Authentication Chain: An alternating sequence of DNS public key (DNSKEY) RRsets and Delegation Signer (DS) RRsets forms a chain of signed data, with each link in the chain vouching for the next. A DNSKEY RR is used to verify the signature covering a DS RR and allows the DS RR to be authenticated. The DS RR contains a hash of another DNSKEY RR, and this new DNSKEY RR is authenticated by matching the hash in the DS RR. This new DNSKEY RR, in turn, authenticates another DNSKEY RRSet and, in turn, some DNSKEY RR in this set may be used to authenticate another DS RR, and so forth until the chain finally ends with a DNSKEY RR whose corresponding private key signs the desired DNS data. For example, the root DNSKEY RRSet can be used to authenticate the DS RRSet for "example." The "example." DS RRSet contains a hash that matches some "example." DNSKEY, and this DNSKEY's corresponding private key signs the "example." DNSKEY RRSet. Private key counterparts of the "example." DNSKEY RRSet sign data records such as "www.example." as well as DS RRs for delegations such as "subzone.example."

Authentication Key: A public key that a DNSSEC-aware resolver has verified and can therefore use to authenticate data. A DNSSEC-aware resolver can obtain authentication keys in three ways. First, the resolver generally is configured to know about at least one public key; this configured data usually is either the public key itself or a hash of the public key as found in the DS RR (see "trust anchor"). Second, the resolver may use an authenticated public key to verify a DS RR and the DNSKEY RR to which the DS RR refers. Third, the resolver may be able to determine that a new public key has been signed by the private key corresponding to another public key that the resolver has verified. Note that the resolver must always be guided by local policy in deciding whether to authenticate a new public key, even if the local policy is simply to authenticate any new public key for which the resolver is able verify the signature.

Authoritative RRSet: Within the context of a particular zone, an RRSet (RRs with the same name, class, and type) is authoritative if and only if the owner name of the RRSet lies within the subset of the name space that is at or below the zone apex and at or above the cuts that separate the zone from its children, if any. RRs of type NSEC, RRSIG and DS are examples of RRSets at a cut that are authoritative at the parent side of the zone cut, and not the delegated child side.

Chain of Trust: See "authentication chain."

Chained Secure Zone: A DNS zone in which there is an authentication chain from the zone to a trust anchor.

Delegation Point: The name at the parental side of a zone cut. That is, the delegation point for "foo.example" would be the foo.example node in the "example" zone (as opposed to the zone apex of the "foo.example" zone). See also "zone apex."

DNS Administrator: Used in this document to cover the person (or persons) tasked with updating zone data and operating an enterprise's DNS server. This term may actually cover several official roles, but these roles are covered by one term here.

DNSSEC-Aware Name Server: An entity acting in the role of a name server that understands the DNS security extensions defined in this document set. In particular, a DNSSEC-aware name server is an entity that receives DNS queries, sends DNS responses, supports the EDNS0 [RFC2671] message size extension and the DO bit [RFC4035], and supports the RR types and message header bits defined in this document set.

DNSSEC-Aware Recursive Name Server: An entity that acts in both the DNSSEC-aware name server and DNSSEC-aware resolver roles. A more cumbersome equivalent phrase would be "a DNSSEC-aware name server that offers recursive service." Also sometimes referred to as a "security-aware caching name server."

DNSSEC-Aware Resolver: An entity acting in the role of a resolver (defined in section 2.4 of [RFC4033]) that understands the DNS security extensions. In particular, a DNSSEC-aware resolver is an entity that sends DNS queries, receives DNS responses, and understands the DNSSEC specification, even if it is incapable of performing validation.

DNSSEC-Aware Stub Resolver: An entity acting in the role of a stub resolver that has an understanding of the DNS security extensions. DNSSEC-aware stub resolvers may be either "validating" or "nonvalidating," depending on whether the stub resolver attempts to verify DNSSEC signatures on its own or trusts a friendly DNSSEC-aware name server to do so. See also "validating stub resolver" and "nonvalidating stub resolver."

Island of Security: A signed, delegated zone that does not have an authentication chain from its delegating parent. That is, there is no DS RR containing a hash of a DNSKEY RR for the island in its delegating parent zone. An island of security is served by DNSSEC-aware name servers and may provide authentication chains to any delegated child zones. Responses from an island of security or its descendents can be authenticated only if its authentication keys can be authenticated by some trusted means out of band from the DNS protocol.

Key Rollover: The process of generating and using a new key (symmetric or asymmetric key pair) to replace one already in use. Rollover is done because a key has been compromised or is vulnerable to compromise as a result of use and age.

Key Signing Key (KSK): An authentication key that corresponds to a private key used to sign one or more other authentication keys for a given zone. Typically, the private key corresponding to a key signing key will sign a zone signing key, which in turn has a corresponding private key that will sign other zone data. See also "zone signing key."

Nonvalidating DNSSEC-Aware Stub Resolver: A DNSSEC-aware stub resolver that trusts one or more DNSSEC-aware recursive name servers to perform most of the tasks discussed in this document set on its behalf. In particular, a nonvalidating DNSSEC-aware stub resolver is an entity that sends DNS queries, receives DNS responses, and is capable of establishing an appropriately secured channel to a DNSSEC-aware recursive name server that will provide these services on behalf of the DNSSEC-aware stub resolver. See also "DNSSEC-aware stub resolver" and "validating DNSSEC-aware stub resolver."

Nonvalidating Stub Resolver: A less tedious term for a nonvalidating DNSSEC-aware stub resolver.

Signed Zone: A zone whose RRsets are signed and which contains properly constructed DNSKEY, Resource Record Signature (RRSIG), Next Secure (NSEC), and (optionally) DS records.

Trust Anchor: A configured DNSKEY RR or DS RR hash of a DNSKEY RR. A validating DNSSEC-aware resolver uses this public key or hash as a starting point for building the authentication chain to a signed DNS response. In general, a validating resolver will need to obtain the initial values of its trust anchors via some secure or trusted means outside the DNS protocol. The presence of a trust anchor also implies that the resolver should expect the zone to which the trust anchor points to be signed. This is sometimes referred to as a "secure entry point."

TSIG Key: A string used to generate the message authentication hash stored in a TSIG RR and used to authenticate an entire DNS message. This is not the same as signing a message, which involves a cryptographic operation.

Unsigned Zone: A zone that is not signed.

Validator: A component that validates DNSSEC signatures. Usually not a separate component but part of a DNSSEC-aware recursive server (sometimes referred to as a validating resolver or validating recursive server).

Zone Apex: The name at the child's side of a zone cut. See also "delegation point."

Zone Signing Key (ZSK): An authentication key that corresponds to a private key used to sign a zone. Typically a zone signing key will be part of the same DNSKEY RRSet as the key signing key whose corresponding private key signs this DNSKEY RRSet, but the zone signing key is used for a slightly different purpose and may differ from the key signing key in other ways, such as validity lifetime. Designating an authentication key as a zone signing key is purely an operational issue: DNSSEC validation does not distinguish between zone signing keys and other DNSSEC authentication keys, and it is possible to use a single key as both a key signing key and a zone signing key. See also "key signing key."

Appendix B—Vendor Specific Steps to Meet Checklist Items

This publication does not attempt to list configuration examples for every DNS component available. The authors have made the decision to only include major open-source implementations to use in the examples. Other implementations may have vendor specific features or configuration options available to achieve the same set of recommendations in this guide. In some cases, implementations may not contain every features described in this publication and so those checklist items would not apply when installing and configuring those implementations.

Whenever possible, a shorthand guide on how to meet the checklist items using a specific implementation will be produced as a separate documents. These documents are not official NIST Special Publications, but supplemental guidance for specific implementations.

The repository for these supplemental documents can be found at **http://www.dnsops.biz/vendors**

Appendix C—Acronyms

Selected acronyms used in the *Secure Domain Name System (DNS) Deployment Guide* are defined below.

A	Address
ACL	Access Control List
AD	Authenticated Data
ANSI	American National Standards Institute
ARP	Address Resolution Protocol
CA	Certificate Authority
ccTLD	Country-code Top-level Domain
CD	Checking Disabled
DHCP	Dynamic Host Configuration Protocol
DNS	Domain Name System
DNSSEC	Domain Name System Security Extensions
DS	Delegation Signer
DSA	Digital Signature Algorithm
DSS	Digital Signature Standard
ECDSA	Elliptic Curve Digital Signature Algorithm
FIPS	Federal Information Processing Standards
FISMA	Federal Information Security Management Act
FQDN	Fully Qualified Domain Name
gTLD	Generic Top-level Domain
HINFO	Host Information
HMAC	Hash-based Message Authentication Code
ICANN	Internet Corporation for Assigned Names and Numbers
IETF	Internet Engineering Task Force
IN	Internet
IP	Internet Protocol
ISP	Internet Service Provider
IT	Information Technology
ITL	Information Technology Laboratory
KSK	Key Signing Key
LAN	Local Area Network
LOC	Location
MAC	Message Authentication Code
MD	Message Digest
MX	Mail Exchanger
NIST	National Institute of Standards and Technology
NS	Name Server

NSEC	Next Secure
NSEC3	Hashed Next Secure
NTP	Network Time Protocol
OMB	Office of Management and Budget
OS	Operating System
PKI	Public Key Infrastructure
PIR	Public Internet Registry
RFC	Request for Comments
RP	Responsible Person
RR	Resource Record
RRSIG	Resource Record Signature
SEP	Secure Entry Point
SHA	Secure Hash Algorithm
SHS	Secure Hash Standard
SOA	Start of Authority
TCP	Transmission Control Protocol
TLD	Top-level Domain
TSIG	Transaction Signature
TTL	Time to Live
TXT	Text
UDP	User Datagram Protocol
ZSK	Zone Signing Key

Appendix D—Bibliography

I. DNS – Resource Records and Generic Architecture

[RFC3833]   D. Atkins and R. Austein, "Threat Analysis of the Domain Name System (DNS)", RFC 3833, August 2004.
[RFC2181]   R. Elz and R. Bush, "Clarifications to the DNS Specification", RFC 2181, July 1997.
[RFC1912]   D. Barr, "Common DNS Operational and Configuration Errors", RFC 1912, Feb. 1996
[RFC1034]   P. Mockapetris, "Domain Names - Concepts and Facilities", STD 13, RFC 1034, November 1987.
[RFC1035]   P. Mockapetris, "Domain Names - Implementation and Specification", STD 13, RFC 1035, November 1987.
[BCP140]   J. Damas and F. Neves, "Preventing Use of Recursive Nameservers in Reflector Attacks", BCP 140, RFC 5358. October 2008.
[RFC2671]   P. Vixie, "Extension Mechanisms for DNS (EDNS0)" RFC 2671 August 1999
[RFC1996]   P. Vixie, "A Mechanism for Prompt Notification of Zone Changes (DNS NOTIFY)" RFC 1996. August 1996.
[IEEE]   S. Rose and A. Nakassis. "Minimizing Information Leakage in the DNS" IEEE Network Magazine vol. 22 no. 2 April 2008.

II. DNSSEC

[RFC4033]   R. Arends, R. Austein, M. Larson, D. Massey, and S. Rose, "DNS Security Introduction and Requirements", RFC 4033, March 2005.
[RFC4034]   R. Arends, et al, "Resource Records for DNS Security Extensions", RFC 4034, March 2005.
[RFC4035]   R. Arends, et al, "Protocol Modifications for the DNS Security Extensions", RFC 4035, March 2005.
[RFC4641]   O. Kolkman and R. Gieben. "DNSSEC Operational Practices", RFC 4641. September 2006.
[RFC5011]   M. StJohns. "Automated Updates for DNS Security (DNSSEC) Trust Anchors". RFC 5011, September 2007.
[RFC5155]   B. Laurie, G. Sisson, R. Arends and D. Blacka. "DNS Security (DNSSEC) Hashed Authenticated Denial of Existance" RFC 5155, March 2008

III. Dynamic Update

[RFC3007]   B. Wellington, "Secure Domain Name System (DNS) Dynamic Update", RFC 3007, November 2000.
[RFC2136]   P. Vixie, S. Thomson, Y. Rekhter, and J. Bound. "Dynamic Updates in the Domain Name System (DNS UPDATE)". RFC 2136, April 1997.

IV. TSIG

[RFC2931]   D. Eastlake 3rd, "DNS Request and Transaction Signatures (SIG(0)s)", RFC 2931, September 2000.
[RFC2845]   P. Vixie, O. Gudmundsson, D. Eastlake 3rd, and B. Wellington, "Secret Key Transaction Authentication for DNS (TSIG)", RFC 2845, May 2000.

[RFC3645]   S. Kwan, P. Garg, J. Gilroy, L. Esibov, J. Westhead and R. Hall. "Generic Security Service Algorithm for Secret Key Transaction Authentication for DNS (GSS-TSIG)". RFC 3645 October 2004.

[RFC4635]   D. Eastlake 3rd, "HMAC SHA TSIG Algorithm Identifiers" RFC 4635, August 2006.

V. Cryptographic Standards and Other USG Guidance Documents

[RFC2104]   H. Krawczyk, M. Bellare, and R. Canetti. "HMAC: Keyed-Hashing for Message Authentication", RFC 2104, February 1997

[FIPS198]   "The Keyed-Hash Message Authentication Code (HMAC)", Federal Information Processing Standard, National Institute of Science and Technology. FIPS 198-1, July 2008.

[FIPS180]   FIPS 180-2, "Secure Hash Standard (SHS)", Federal Information Processing Standard, National Institute of Science and Technology. FIPS 180-2, August 2002.

[FIPS186]   "Digital Signature Standard" FIPS 186-3. March 2008.

[800-57P1]   E. Barker, W. Barker, W. Burr, W. Polk and M. Smid. "Recommendation for Key Management – Part 1: General" NIST Special Publication 800-57 Part 1 (Revision 3). July 2012.

[800-57P3]   A. Clay, E. Barker, W. Burr, W. Polk, S. Rose. "Recommendation for Key Management – Part 3: Application-Specific Key Management Guidance" NIST Special Publication 800-57 Part 3, December 2009

VI. Security Vulnerabilities

[CERT]   CERT®/CC's Vulnerability Notes Database at http://www.kb.cert.org/vuls/

[NVD]   NIST NVD metabase at http://nvd.nist.gov/

[BINDSEC]   BIND vulnerabilities page at http://www.isc.org/products/BIND/bind-security.html

www.ingramcontent.com/pod-product-compliance
Lightning Source LLC
Chambersburg PA
CBHW081459170526
45166CB00008B/2482